No Postage
Necessary
If Mailed
in the
United States

BUSINESS REPLY MAIL
FIRST CLASS MAIL PERMIT NO. 26 NEW YORK, NY

Postage will be paid by addressee

McGraw-Hill, Inc.
Attn: Janet Tafrow
11 West 19th Street
New York, NY 10114-0144

1991 Yearbook Supplement to McGraw-Hill's NATIONAL ELECTRICAL CODE® HANDBOOK

Joseph F. McPartland
Editorial Director/Publisher
edi (Electrical Design and Installation) *Magazine*
452 Hudson Terrace
Englewood Cliffs, NJ 07632

Brian J. McPartland
Associate Publisher/Editor
edi (Electrical Design and Installation) *Magazine*
452 Hudson Terrace
Englewood Cliffs, NJ 07632

ASSISTANT EDITORS

William C. Broderick, P.E.
Electrical Consultant, River Edge, NJ

Brendan A. McPartland
Facilities Engineer
Coldwell Banker, White Plains, NY

Jack E. Pullizzi
Electrical Construction Specialist
AT&T Bell Labs, Holmdel, NJ

Gilbert L. Thompson
Chief Electrical Inspector
Baltimore County, Baltimore, MD

McGraw-Hill, Inc.

New York St. Louis San Francisco Auckland Bogotá
Caracas Hamburg Lisbon London Madrid
Mexico Milan Montreal New Delhi Paris
San Juan São Paulo Singapore
Sydney Tokyo Toronto

1 2 3 4 5 6 7 8 9 0 DOC/DOC 9 7 6 5 4 3 2 1

ISBN 0-07-045904-5 {HC}
ISBN 0-07-045905-3 {PBK}

The sponsoring editor for this book was Harold B. Crawford, the editing supervisor was Alfred Bernardi, and the production supervisor was Suzanne W. Babeuf. It was set in Century Schoolbook by McGraw-Hill's Professional Book Group composition unit.

Printed and bound by R. R. Donnelley & Sons Company.

Contents

An Invitation from the Authors

The *1991 Yearbook Supplement to McGRAW-HILL's NATIONAL ELECTRICAL CODE® HANDBOOK* was developed to help you resolve the many Code-related issues that electrical professionals face on a daily basis. You the reader bring to this *Yearbook Supplement* valuable personal experiences in working with the Code that can be shared with your colleagues. We therefore invite you to describe in letter form ways in which you have resolved problems in applying Code requirements. We also invite you to ask questions regarding Code interpretation that have been troublesome. Such questions will be answered in future *Yearbook Supplements*.

Letters are to include your full name, address, and affiliation, which will be withheld on request should your letter be reprinted in the *Yearbook Supplement*. Such letters become the property of McGraw-Hill.

ABOUT THE AUTHORS

JOSEPH F. MCPARTLAND was the Editorial Director of *Electrical Construction and Maintenance, Electrical Construction and Maintenance Products Yearbook, Electrical Wholesaling,* and *Electrical Marketing Newsletter*. He is now Editorial Director/Publisher of *edi* (*Electrical Design and Installation*) Magazine. He has authored twenty-six books on electrical design, electrical construction methods, electrical equipment, and the *National Electrical Code®*. For over 38 years he has been traveling throughout the United States conducting seminars and courses on the many aspects of electrical design, engineering, and construction technology for electrical contractors, consulting engineers, plant electrical people, and electrical inspectors. Mr. McPartland received a Bachelor of Science degree in Electrical Engineering from Thayer School of Engineering, Dartmouth.

BRIAN J. MCPARTLAND was an associate editor of *Electrical Construction and Maintenance* and has more than 14 years' experience in electrical technology. He is now Associate Publisher/Editor of *edi* (*Electrical Design and Installation*) Magazine. After serving in the U.S. Navy Submarine Force, during which time he earned his degree in electrical technology, Brian held positions in both product engineering and sales with various electrical equipment manufacturers. He is coeditor of McGraw-Hill's *Handbook of Practical Electrical Design* and coauthor of *McGraw-Hill's National Electrical Code® Handbook* and EC&M's *Illustrated Changes in the 1990 National Electrical Code®*.

Liquidtight and Standard Flexible Metal Conduit MUST Be "Listed"

Sec. 110-2. The intent of the NEC in this section is to place a strong insistence on third-party certification of the essential safety of equipment and component products used to assemble an electrical system.

Over the past few years there has been a substantial increase in the use of liquidtight and standard flexible metal conduit, which has not been certified by a third-party testing company and "listed" as either meeting the appropriate standards or having been tested and found suitable for the application. These products, the so-called "extra flexible," or Type EF, as it is sometimes known, has been applied in residences, commercial buildings, and industrial facilities alike in violation of the intent of the NEC, as well as the letter of the Occupational Safety and Health Administration's (OSHA's) rules on the use of "listed," "labeled," or otherwise "certified" equipment.

The main reason for the proliferation of nonlisted liquidtight and standard flexible metal conduit is the cost differential. As can be expected, the nonlisted version of these products costs less than the properly listed and labeled item. Both products are available through most electrical wholesale houses and well-meaning distributor salespeople frequently offer the nonlisted product to a contractor as a less expensive alternative to the listed liquidtight and standard flexible metal conduit.

DO NOT accept that argument *or* the nonlisted product (Figure 1). Verdicts rendered in recent court cases have consistently held that the use of the nonlisted liquidtight or standard flexible metal conduit is a violation. The contractor, not the distributor or manufacturer, was the one held responsible for the accident or incident that resulted.

In Sec. 110-2, the NEC regulates the use of electrical equipment and products. Only equipment that has been "approved" may be used. The meaning of the word "approved" must be taken to be that as given in Article 100 of the NEC, which is "acceptable to the authority having jurisdiction." Although the electrical inspector in a given jurisdiction, who is the final judge of acceptability in accordance with Sec. 90-4, is not *required* to use "listing" or "labeling" by a nationally recognized testing lab as the deciding factor in determining the suitability of a product, inspectors in the vast majority of jurisdictions typically re-

Figure 1 Bootleg liquidtight flex. In addition to being improperly supported in this application, use of this contraband (nonlisted liquidtight flexible metal conduit, which is a material illegally used in the trade) violated the *intent* of the NEC rules and the *letter* of OSHA's rules dealing with the use of "listed," "labeled," "accepted," or otherwise "certified" equipment when it's available. In this photo, the absence of a clamp from the more-than-3-ft length of this conduit below the one clamp is a clear violation of Sec. 351-8, which only permits an unclamped length of more than 3 ft when using the flex to supply a lighting fixture, as covered in Exception No. 3 of Sec. 351-8. In addition, the installer failed to provide a clamp within 12 in. of the A/C unit's termination point.

quire such certification. While the NEC is nonspecific with respect to *requiring* third-party certification, the rules of OSHA, which apply to virtually all commercial, industrial, and institutional buildings, as well as to certain residences, are very rigid in insisting on product certification.

The clear effect of OSHA regulations is to require that "listed," "labeled," "accepted," or "certified" equipment be used whenever available. That is, if *any* nationally recognized testing lab "lists, labels, accepts, or certifies" an electrical system component, then the listed, labeled, accepted, or certified product *must be* used to comply with OSHA regulations. Because liquidtight and standard flexible metal conduit are products that are "listed" and "labeled" by a nationally recognized testing lab, it is clearly a violation of OSHA rules to use a nonlisted version of liquidtight or standard flexible metal conduit.

The only time a nonlisted, nonlabeled, or noncertified component may be used is if it is "of a kind" that *no* nationally recognized testing lab covers. For example, there is no liquidtight flexible metal conduit listed for use in an elevated ambient. According to the listing instructions, liquidtight flexible metal conduit investigated by Underwriters Laboratories "is intended for use in wet locations or where exposed to mineral oil, both at a maximum temperature of 60°C." Because Sec. 110-3(b) requires listed or labeled equipment to be used in accordance with the listing and labeling instructions, liquidtight flexible metal conduit that is listed and labeled by Underwriters Laboratories must *not* be used in applications where the temperature exceeds 60°C. In such applications, only nonlisted "high-temperature" liquidtight flexible metal conduit would be acceptable.

Always insist that your distributor supply you with listed and labeled liquidtight and standard flexible metal conduit. Don't let fear of competition from other electrical contractors or the opportunity to "save" a couple of bucks cloud your vision. The "extra" few dollars spent should be passed along to the customer, who is the actual beneficiary of such practice. Additionally, constant insistence on third-party certification is cheap insurance against a lawsuit that could destroy your business.

Protect Outdoor Electrical Equipment

Secs. 110-3(a), 110-17(b), 230-50(a), 230-70(c), and 240-24(c). **Outdoor electrical equipment exposed to physical damage—such as impact by car or truck—MUST be provided with substantial protection.**

As shown in Figure 1, service equipment for a building is commonly installed outdoors in places where the equipment might be struck by a moving car or truck—such as adjacent to a driveway. Because of the importance of electrical supply continuity for modern buildings, most electrical inspectors are concerned about minimizing interruption of

Figure 1 Outdoor electrical equipment exposed to physical damage requires substantial protection. The concrete-filled poles (arrow) embedded in underground concrete bases provide such protection for the service equipment shown. Although it may often be a controversial matter to relate this requirement to one specific Code rule, several rules have some bearing on the situation and can be applied by an alert electrical inspector.

building power by providing adequate protection of the service equipment against physical damage.

In Figure 1, two lengths of concrete-filled pipe (arrow) are embedded in concrete underground bases and positioned to protect the enclosure housing the meters and the service disconnects (circuit breakers) supplying three store units in a one-level multioccupancy commercial building. Although such protection is logical and extremely important to prevent costly damage to the equipment—as well as assuring continuity of service to the stores—it is often a controversial matter to relate this to specific Code rules, even though a number of rules bear on this concern, as discussed in the following paragraphs.

To begin with, Sec. 110-3 has a heading that includes the phrase "Installation and Use of Equipment" and subpart (a) (2) requires equipment to have "Mechanical strength and durability, including, for parts designed to enclose and protect other equipment, the adequacy of the protection thus provided." The two posts in Figure 1 protect against the **"inadequacy"** of the protection provided simply by the enclosure of the service equipment.

Section 110-17(b) is entitled "Prevent Physical Damage" and says that "In locations where electric equipment would be exposed to physical damage, enclosures **or guards** shall be **so arranged** and **of such strength** as to prevent such damage." That is a very clear, specific, and simple rule that could be used to require the posts shown in Figure 1, with the phrase **"of such strength"** taken to require "guards" (the posts) because the meter enclosure could not provide the protection against vehicle damage.

Because the conductors coming into the top of the service enclosure are service conductors, Sec. 230-50(a) would apply, and that rule requires service conductors "in exposed places **near driveways**" to be protected by rigid or intermediate metal conduit or "(5) by other approved means."

Section 230-70(c) requires service disconnecting means to be "suitable for the prevailing conditions," which could be taken to require supplemental protection against physical damage where needed.

The circuit breaker disconnects in the service enclosure in Figure 1 must satisfy Sec. 240-24(c): "Overcurrent devices shall be located where they will not be exposed to physical damage"—such as vehicle impact.

Equipment and Conductor Temperature Limitations

Sec. 110-3(b). Beware of conductor and equipment temperature limitations!

The primary enemy to virtually all electrical equipment and conductors is the I^2R heat generated during use. Although this heating of equipment and conductors is normal and predictable, if the amount of heating is not limited, it can adversely affect the insulating material used, and result in premature failure of the equipment or conductors. As a result, when equipment or conductors are evaluated by UL, or other third party testing labs, for listing, the insulation system used is also examined to determine and establish the maximum temperature at which the equipment or conductor can be safely operated without causing damage to the insulation.

There are, however, many installations throughout the country where temperature limitations of equipment and conductors are not properly coordinated and the equipment and/or conductors are actually overloaded.

It is important to note that the magnitude of overload only influences the amount of time before failure occurs. That is, a slight overload will take a long time to cause insulation breakdown and failure and a great overload will cause insulation breakdown and failure more quickly. But in either case, the equipment and conductors will not have as long a service life as would normally be expected.

In addition to a shortened service life, failure to observe the temperature limitations of equipment and conductors is a violation of the NEC. In Sec. 110-3(b), the NEC requires all listed and labeled products to be used in accordance with their listing or labeling instructions. This requirement has been repeatedly interpreted—by inspectors in the field, as well as judges in the courts—to mandate strict adherence to all instructions given by the listing lab. And among the many thousands of installation details given in the UL's "General Information Directory" or "White Book," as it is often called, there are general and specific temperature limitations that must be observed when using listed equipment and conductors.

The following discussion will illuminate the meaning and application of the general temperature limitations to provide a basis for assuring Code compliance and maximum service life of equipment and conductors.

General instructions

On pages 6, 7, and 8 of the 1990 edition of the UL's White Book, a variety of "general" instructions are given. Most important to this discussion are the instructions given for "Appliances and Utilization Equipment Terminations" and "Distribution and Control Equipment Terminations." Unless a specific equipment listing instruction states otherwise, these two provisions establish maximum temperature ratings for terminations in *all* equipment that is to be interconnected using Code-recognized conductors and cables to form the building's or facility's electrical system.

Although these two provisions address the same concern—temperature limitations of terminations—each defines the limitation in a slightly different way. However, the net effect is nearly the same.

Appliances and utilization equipment

Under "Appliances and Utilization Equipment," the White Book states that, unless otherwise marked, all field terminations in such equipment are based on the use of 60°C-insulated conductors for circuits rated at 100A or less. And for circuits rated over 100A, the termination provisions are based on the use of 75°C-insulated conductors. These statements are intended to indicate that, depending on the rating of the supply circuit, the terminations within the equipment provided by the manufacturer for connection to branch-circuit conductors are evaluated using either a 60°C- or 75°C-rated conductor at their 60°C- or 75°C-ampacity, as given in Table 310-16. Additionally, UL is indicating that when the appliance or utilization equipment is installed, unless the equipment is marked otherwise, the ampacity of the conductors for supply circuits rated 100A or less is to be based on the ampacity given for a conductor of the same size with 60°C-rated insulation, as shown in Table 310-16. And for conductors of supply circuits rated over 100A, the ampacity is to be based on the ampacity given for the 75°C-rated insulations, as shown in Table 310-16 for a given conductor size.

Notice that the UL statement relates this requirement to the "circuit rating" and not the "nameplate rating." Because UL takes requirements of the NEC into consideration when evaluating appliances or utilization equipment, if a piece of equipment is to be supplied by a circuit that is required by an NEC rule to be rated at a greater ampere value than the equipment's nameplate ampere rating, the equipment will be evaluated accordingly. That is, when a piece of equipment has a nameplate rating that is 100A or less, and the NEC would require a circuit rated over 100A to supply the equipment, the equipment would be evaluated using 75°C insulated conductors at their full 75°C ampacity. And, therefore, when installed, such equipment would be permitted to be supplied by 75°C insulated conductors at the full 75°C ampacity.

An additional statement makes clear that higher temperature-rated insulations may be used provided the conductor ampacity is based on the 60°C value for circuits rated up to 100A and the 75°C value for circuits over 100A. That is, the amount of current carried by a conductor with a higher temperature-rated insulation must not be greater than the value of current shown in Table 310-16 for the same size conductor with either 60°C- or 75°C-rated insulation, depending on the rating of the supply circuit.

Application of these requirements would show that when the NEC requires a circuit rated at 100A to supply an appliance or piece of utilization equipment, and the appliance or utilization equipment is not marked with a specific temperature rating or temperature-rating and conductor size, conductors connected to such equipment would have to be at least a No. 1 AWG, copper with 60°C-rated insulation. Additionally, conductors with either 75°C- or 90°C-rated insulations may also be used, *BUT,* where 75°C-rated (e.g. THW) or 90°C-rated (e.g. THHN) insulation is used, the ampacity must be no more than that given for a 60°C-rated insulated wire. Therefore, in this case, the THW- or THHN-insulated conductors must carry not more than 110A, after derating for an elevated ambient and/or number of conductors, as may be necessary.

If conductors are to be connected to an appliance or utilization equipment where the supply circuit would be required to be rated at more than 100A, and there is no specific temperature rating or temperature rating and conductor size marked on the equipment, the UL information tells us that conductors with 75°C-rated insulation may be used at the full 75°C ampacity shown in Table 310-16. Conductors with 90°C-rated insulations (RHH, THHN, XHHW, etc.) may be used, *BUT* the ampacity of these conductors must be taken to be that of a 75°C insulated conductor of the same AWG- or kcmil-size, after any derating for ambient or number of conductors is applied. Additionally, if the equipment is not marked with a specific temperature rating or a temperature rating and conductor size, then it would also be permissible to use a conductor with 60°C-rated insulation, provided the conductor selected had sufficient ampacity as described within applicable Code rules for the load being supplied.

Guidance is also provided for equipment whose termination provisions are based on the use of conductors with higher temperature-rated insulation. When an appliance or piece of utilization equipment is intended for use with conductors of higher temperature ratings than described above, such equipment will be marked to show the temperature rating and size of conductor necessary or, just a temperature rating.

If the equipment is marked to show only a higher temperature rating, then the insulation of the conductor used must be rated as indicated, *BUT* the conductor ampacity must be not more than the value given in Table 310-16 for the same size conductor with 60°C-rated insulation for applications up to 100A or with 75°C-rated insulation for applications over 100A. Additionally, any derating required may be applied against the ampacity value given for the higher temperature-rated insulation in Table 310-16, but the final value of ampacity must not exceed that shown for a 60°C- or 75°C-insulated conductor of the same size, where the equipment is supplied by a circuit rated 100A or less, or over 100A, respectively.

If the equipment is marked with both a temperature rating and a conductor size, only a conductor of the indicated size and insulation temperature-rating may be used.

Distribution and control equipment

As indicated above, the UL temperature limitations for distribution and control equipment are defined on a different basis than those for appliances and utilization equipment. Instead of considering the ampere rating of the supply circuit conductors, for distribution and control equipment, temperature limitations of the terminations are based on the size of the circuit conductors used, but the net effect is almost identical.

The general guideline for terminations within distribution and control equipment states that, unless marked otherwise, when the terminations are intended for use with conductors sized from No. 14 to No. 1 AWG, the ampacity must be based on the ampacity given in Table 310-16 for a conductor of the same size with 60°C-rated insulation. And when terminating devices intended are intended for use with conductors Nos. 1/0 AWG and larger, the ampacity of the conductor used must be based on the 75°C value for a conductor of the same size, as shown in Table 310-16. With this differentiation in mind, it can be seen in Table 310-16 that the ampere value at the point where the UL changes to the higher temperature rated conductors when evaluating such equipment is about 100A, which is the value used as the dividing line between 60°C- and 75°C-rated conductors supplying appliance and utilization equipment.

The White Book also states that a marking of "75C only" or "60/75C" on a switch, breaker, or lug does not necessarily mean that 75°C-insulated conductors may be used at their full 75°C ampacity. In

Listed equipment must be used in accordance with all instructions given by the third-party testing lab. Even though this lug is marked AL9CU—which indicates it is for use with either copper or aluminum conductors at their full 90°C ampacity (arrow)—a general instruction given in the UL's "General Information Directory," the so-called "White Book," states that such a lug may be used with 90°C-rated conductors at their full 90°C ampacity, *only* if the enclosure or equipment is also marked to indicate that 90°C-rated conductors may be used at their full 90°C ampacity.

Distribution and control equipment intended for use with conductors in sizes from No. 14 to No. 1 AWG are to be used with 60°C-rated conductors at their 60°C ampacities, unless otherwise marked. The label on this CB states: "60/75°C WIRE." If this CB is mounted in an individual enclosure or, if it is mounted in a panelboard that is also marked for use with 75°C-rated wires, then 75°C-rated conductors may be used at the full 75°C ampacity.

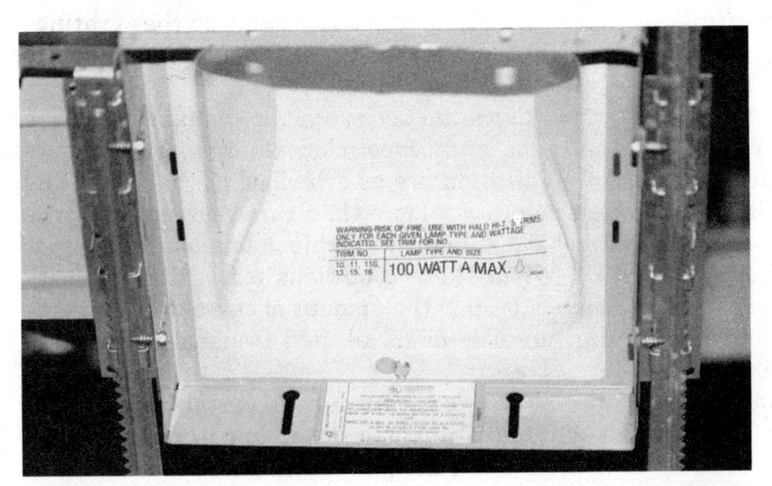

Appliances and utilization equipment supplied by circuits rated 100A or less are evaluated using 60°C-rated conductors, unless the appliance or equipment is otherwise marked. Additional restrictions will also be given either on labels or in the manufacturer's installation instructions. The label inside this fixture warns that the fixture is suitable for use with an incandescent lamp rated at not more than 100W. Use of any lamp that is rated over 100W would be in violation of the manufacturer's labeling instruction, and would therefore be a violation of Sec. 110-3(b) in the National Electric Code.

addition to such marking on the switch, CB, or lug, the enclosure or equipment in which the switch, breaker, or lug is installed must also be so marked (i.e. "75C only" or "60/75C"), independently of the marking on the switch, CB, or lug.

While this is the general requirement, an additional statement clarifies that when a switch or CB is to be used by itself, such as in a separate enclosure, if the switch or breaker is marked for use with 75°C-insulated conductors, then the enclosure is not required to be independently marked, and it would be permissible to use such conductors at their full 75°C ampacity. THIS PERMISSION APPLIES TO SWITCHES AND CBs INSTALLED IN INDIVIDUAL ENCLOSURES, ONLY! For lugs, only when the enclosure or equipment is also marked is it permissible to use 75°C-insulated conductors at their full 75°C.

The UL general listing instruction for distribution and control equipment also states that higher temperature-rated conductors may be used, provided that the conductor's ampacity is not greater than that given in the 60°C column for sizes No. 14 to No. 1 AWG, and the 75°C column for No. 1/0 AWG and larger, as given in Table 310-16.

Conductors

As we have seen, these UL requirements are all based on the use of conductors with an ampacity not more than that shown for certain insulation ratings as given in Table 310-16. But what do these ratings actually mean and how are they related to the overheating of equipment terminations?

Table 310-16 in the NEC gives the Code-recognized ampacities for a variety of copper, aluminum, and copper-clad aluminum conductors when the conditions of application are as described at the top of this table. That is, where there are not more than three current-carrying conductors in raceway or cable and the ambient temperature is not more than 86°F (30°C). When these conditions exist, the ampacity value given within Table 310-16 is the amount of current in amperes that a conductor of a given size, material, and insulation can carry continuously.

As can be seen in Table 310-16, the amount of current that a conductor of a given material and size is permitted to carry varies with the type of insulation used. For example, where there are not more than three current-carrying conductors in a raceway or cable and the ambient temperature is not over 86°F (30°C), the ampacity of a No. 1 AWG, copper conductor with TW insulation is 110A. Under the same conditions of application, a No. 1 AWG, copper conductor with THW insulation would have an ampacity of 130A. And a No. 1 AWG, copper

In addition to temperature ratings, equipment labels and instructions will give the designer and installer a variety of requirements regarding the application and installation of the listed equipment. Always, take the time to review these instructions and labels to assure that the equipment is being used in accordance with the manufacturer's listing instructions.

conductor with THHN would have an ampacity of 150A under the same conditions of use.

The reason for the difference in ampacity for the different insulation types has to do with the maximum temperature that the various insulations can withstand. With TW-insulated wire, the maximum acceptable operating temperature is 60°C. For THW insulation, the maximum operating temperature is 75°C. And THHN has a maximum operating temperature of 90°C ampacity.

When the conditions of use are as described in the heading to Table 310-16, the values given in Table 310-16 represent that amount of current that will cause the copper, where it is in contact with the insulation, to reach and stabilize at the temperature indicated for the particular insulation type. That is, for a No. 1 AWG, copper conductor with Type TW insulation, where there are not more than three current-carrying conductors in a raceway and the ambient temperature is not over 86°F, 110A is the amount of current that will cause the copper to reach and stabilize at a temperature of 60°C. Any additional current will cause the copper to "run hotter" than the 60°C and will contribute to insulation breakdown. The same is true for the 75°C- and the 90°C-rated insulations.

When using conductors to interconnect the various pieces of distribution and control equipment, as well as appliance and utilization equipment, the UL rules discussed above warn that the terminations

As is the case with many conductors manufactured today, this cable is marked THHN or THWN. This marking is intended to indicate that when used in a dry location, the ampacity of this conductor may be taken to be that of a THHN-insulated (90°C) conductor, as given in Table 310-16. But, when used in a wet location, as defined by the NEC, the ampacity of this conductor must be based on that given in the 75°C column of Table 310-16, which gives the ampacity for THWN-insulated conductors. Regardless of whether this conductor is used in a dry or wet location, the temperature rating of the equipment terminations must be carefully correlated with the conductor ampacity to prevent overheating and premature failure.

within such equipment were evaluated using conductors that were operated at a specific temperature. Because these terminations were tested at a specific temperature, proper operation and normal service life can be expected only if the equipment is used as it was tested. And, failure to do so is to violate the equipment's listing instruction.

For example, with a panelboard whose main lugs or CB are intended for use with No. 1 AWG conductors, unless the terminating device and enclosure are otherwise marked, the No. 1 AWG conductors must have an ampacity no greater than would be permitted in the 60°C column of Table 310-16 for a No. 1 AWG conductor. Use of 75°C- or 90°C-insulated wire is permitted, but the ampacity or amount of current—which is proportional to the amount of heat generated—must not be greater than the ampere value given for conductors of the same size with 60°C-rated insulations.

In addition to violating the listing instructions and, therefore, Sec. 110-3(b), use of 75°C- or 90°C-rated wire at their full 75°C or 90°C ampacities will cause the terminations to run hot. This is completely undesirable as it can contribute to equipment failure and such failure may be violent and destructive in nature. It must also be remembered that if the terminations within the panelboard were evaluated at a 60°C maximum, the amount of wiring space in the enclosure, which is related to the enclosure's ability to dissipate heat, was also evaluated on the basis of a 60°C maximum. Higher-rated conductors used at their higher ampacities can also become overloaded because the enclosure, which was evaluated using conductors operating at 60°C, may not have sufficient capacity to permit proper dissipation of the additional heat caused by the additional current. This is an important point. Although conductors with higher-rated insulations may still have any derating for number of conductors and/or elevated ambient applied against their higher ampere ratings, as given in Table 310-16, if the final value of ampacity is greater than that shown for the same size conductor in either the 60°C or 75°C column, depending on the conductor size, this "additional" current will generate "additional" heat. Remember, this "additional" heating, no matter how slight, can eventually cause the conductor insulating material to break down and result in premature conductor failure.

In addition to these general instructions regarding temperature-limitations, there are other limitations described in the general instructions, as well as many specific listing instructions dealing with the installation and use of listed equipment. Take the time to read and understand every such instruction before designing systems with, or installing listed equipment and conductors to ensure strict adherence to all listing instructions given by the third party testing lab. Remember, compliance with these requirements is mandatory!

Torquing of Electrical Terminals

First, there was Sec. 110-3(b), and now there's Sec. 430-9(c).
Gradually it has become a requirement of the NEC that electrical terminals to which circuit conductors are connected must be tightened to a specific prescribed value of contact pressure by means of a torque screwdriver or torque wrench. From field investigations and extensive laboratory testing and evaluations by manufacturers, NEMA, UL, the National Bureau of Standards, and others, it has been made perfectly clear that joints and terminations in the path of electric current flow are the weakest links in electrical systems. Ineffective terminations—made with either inadequate or excessive tightening force—account for the vast majority of electrical failures, causing unnecessary and often very expensive destruction of electrical equipment as well as creating proven safety and fire hazards.

But, for all the years that those facts have been widely known and have plagued the industry and the public at large, the whole matter of field connections of electrical conductors has been completely exempt from rules of the National Electric Code and left entirely to the discretion (or the "mercy") of the field electrician. Although proposals have been submitted over past years to include mandatory NEC rules on this critically important safety matter, and although individual design and construction organizations have made torquing mandatory for work covered within their jurisdictions, there has not been any nationally applicable standard rule on field-made electrical joints and terminals. Of course, electrical joints and terminations made during manufacturing and assembly of equipment have long been subject to very specific torquing requirements, which are viewed as absolutely indispensable in assuring safe, long-lasting operation and minimizing product liability exposure.

Now, however, a number of factors combine to place torque tightening of electrical terminals within clear requirement of NEC rules. In the 1987 editions of the Underwriters Laboratories' *Electrical Construction Materials Directory* (Green Book) and *General Information* (White Book), there are requirements for circuit breakers and enclosed switches which state: "All equipment manufactured after October 1, 1986 will be marked to show a tightening torque for all wire connectors intended for use with field wiring." That rule directly interacts with NEC Sec. 110-3(b), which *requires* (a mandatory rule) all listed and labeled products to be used strictly in accordance with any instructions that are given in a testing laboratory's listing book (such as the UL Green Book or White Book). To satisfy the NEC, a torque screwdriver or torque wrench will have to be used to assure that the conductors connected to breaker and switch terminals have the prescribed terminal pressure. That will be a logical follow-up to the fact that manufacturers are already publishing "recommended" torque values in catalogs and specification sheets for distribution equipment and on boxes in which lugs and connectors are packaged (Figure 1). And, there has been an indication from UL that the requirement on torquing of electrical terminals will be moved up to the "General Information" section at the front of the directories to make torquing a requirement for all field connections to a much wider range of electrical equipment that is so marked.

It is also expected that the NEC will also adopt very specific and comprehensive rules in future editions to implement more fully the requirement that derives from the above-described position. The concept of torquing for maximum effectiveness of electrical connections will go beyond just circuit breakers and switches and apply to all electrical connections made in the field. Certainly, the clear indication that universal torquing is today a reality suggests that now is the time for all electrical designers and installers to develop and deepen their knowledge about this subject.

As evidence of more specific NEC rules is Sec. 430-9(c) of the 1987 edition, which requires that screw terminals of all control-circuit devices (push buttons, float switches, pressure switches, limit switches, etc.) used with the coil circuits of magnetic motor starters wired with No. 14 or smaller copper conductors *must* be torqued to a contact pressure of 7 pound-inches (lb-in.)—unless the device is marked for a different torque value (Figure 2).

Important Note: The wording of Sec. 430-9(c) calls for such terminals to be "torqued to a minimum of 7 pound-inches." That conveys the erroneous concept that any value of torque over 7 lb-in. would be

RECOMMENDED TORQUE VALUES—
FOR MAIN BREAKER OR MAIN LUG

LOADCENTERS THRU 125 AMPS		LOADCENTERS 150-225 AMPS	
No. 14-8 AWG	20 in.-lbs.	No. 3-1 AWG	125 in.-lbs.
No. 6-4 AWG	35 in.-lbs.	No. 1/0-2/0 AWG	150 in.-lbs.
No. 3-2/0 AWG	50 in.-lbs.	No. 3/0-4/0 AWG	200 in.-lbs.
		250-450MCM	250 in.-lbs.
		500-750MCM	300 in.-lbs.

Figure 1 Typical label used by equipment manufacturers to indicate "recommended" values of torque-tightness for terminal connections in panelboards presents data that ought to be taken as *mandatory*.

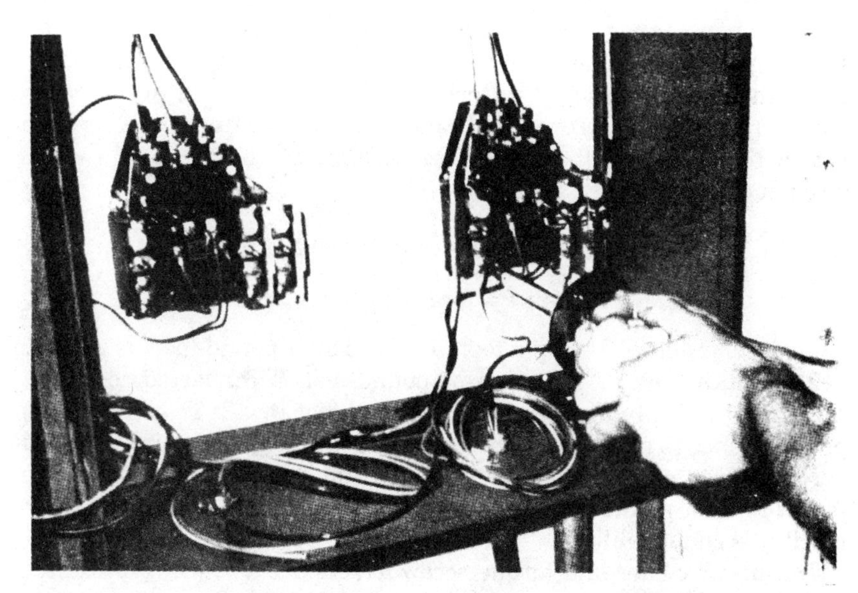

Figure 2 NEC Sec. 430-9 requires torquing to a value of 7 pound-inches for *all* screw terminals of "control circuit devices" of motor controllers—such as by means of a torque screwdriver being used on control-circuit connections to a motor starter.

proper. Although a torque value of, say, 14 lb-in. would satisfy the literal requirement of that rule, excessive torque has been found to be at least as objectionable as inadequate torque. For each type and size of wire in each type of connector assembly, there is only one value (or a narrow range of values) of torque that will provide a low-resistance connection of long-term reliability. The use of a torque screwdriver or torque wrench is intended to assure tightening of any terminal to that very precise value of optimum contact pressure—and not just to exceed a minimum value of tightness. In Sec. 430-9(c), the rule should simply call for an indication of 7 lb-in. on the calibrated pointer-scale of the torque screwdriver or torque wrench.

Figure 3 represents the basic concept behind torquing of a threaded bolt or screw of a connector. Torque is the amount of tightness of the

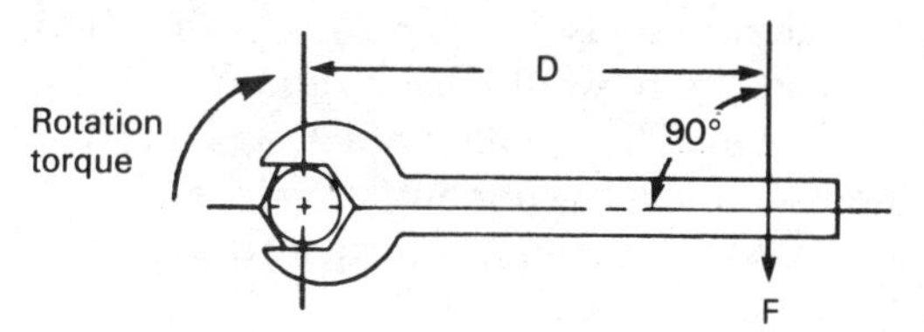

Figure 3 Torque is rotational force and is equal to "force" (F, here) in pounds times "distance" (D, here) in either inches or feet and is described as either "pound-inches" or "pound-feet."

screw or bolt in the threaded hole; that is, torque is the measure of the twisting movement that produces rotations around an axis. Such turning tightness is measured in terms of the force applied to the device that is rotating the screw or bolt and the distance from the axis of rotation to the point where the force is applied to the handle of the wrench or screwdriver:

$$\text{Torque (lb-ft)} = \text{force (lb)} \times \text{distance (ft)}$$

$$\text{Torque (lb-in.)} = \text{force (lb)} \times \text{distance (in.)}$$

Because there are 12 in. in a foot, a torque of "1 pound-foot" is equal to "12 pound-inches." Any value of pound-feet is converted to pound-inches by multiplying the value of pound-feet by 12. To convert from pound-inches to pound-feet, the value of pound-inches is divided by 12.

Note: The expressions "pound-feet" and "pound-inches" are preferred to "foot-pounds" or "inch-pounds," although the expressions are used interchangeably.

Torque wrenches and torque screwdrivers are designed, calibrated, and marked to show the torque (or turning force) being exerted at any position of the turning screw or bolt.

A typical torque screwdriver has a torquing range of 0 to 25 lb-in. The torque screwdriver has changeable tips for use on slotted screws, Phillips-head screws, hex-head bolts, and Allen-head bolts. Such a tool does not generally require periodic zero-setting of the pointer, but provision is made for resetting zero if needed.

For torquing terminals used with conductors larger than No. 10, the usual specified torque values exceed the 25-lb-in. capacity of the torque screwdriver, and a torque wrench must be used. There are two different types of torque wrenches—the ratchet type and the beam type (Figure 4). A typical ratchet torque wrench has adjustable torque setting from 5- to 150-lb-in. capacity. A concentric sleeve on the handle provides for micrometer adjustment of desired torque, with major scale gradations of 10 lb-in. and minor scale gradations of 1 lb-in.

The direct-reading beam-type torque wrench is commonly used to tighten terminals for larger conductors, such as 500-MCM copper conductors, to a torque of 300 lb-in. Because this wrench has a longer handle than the socket wrench, it can apply a given torque with less force required at the handle end than would be required for the shorter handle on the ratchet wrench. This beam wrench has a range from 0 to 600 lb-in. and can be used for torquing any terminations up to that value and for all terminations that require torque in excess of the 150-lb-in. capacity of the ratchet wrench.

After any terminal is tightened to its prescribed torque setting, the terminal should not be retightened to that same value at a later date.

Figure 4 Torquing tools—(left to right) torque screwdriver, ratchet-type torque wrench, and beam-type torque wrench—must now be used in making electrical connections to equipment terminals to assure compliance with *mandatory* rules of the NEC and Underwriters Laboratories.

Electrical connections properly made and torqued at the time of installation do not require periodic retightening. Once set, a termination should be left alone unless and until any indication of malfunction is detected—such as overheating revealed by a thermographic (infrared-heat) inspection of the terminals. Figure 5 shows a characteristic of electrical connections that should be carefully understood.

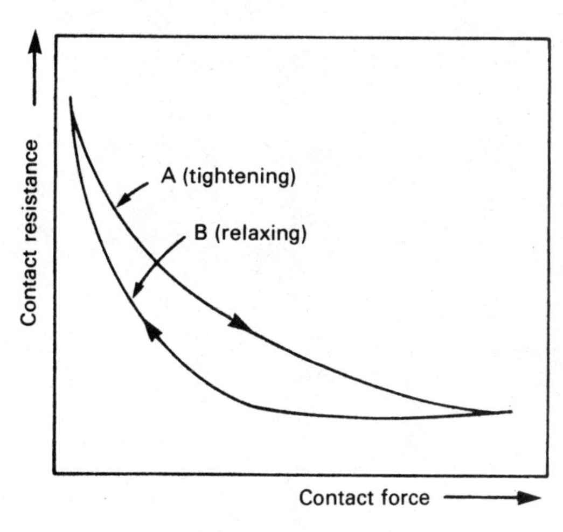

Figure 5 In a properly tightened terminal, constant contact resistance is maintained even though relaxation of the conductor metal will cause reduction of the force of contact (the torque value) at the terminal. Such reduction must not be assumed to dictate periodic retightening.

Curve A shows how tightening of a connector or a conductor produces increasing contact force (going left to right) and a decreasing contact resistance (coordinate on left). That action should be applied up to the specified value of torque. Then, with the passage of time, the conductor material under pressure from the connector relaxes somewhat because of molecular movement in the conductor material and stabilizes at some value of contact force lower than the original applied force. Curve B shows the reduction in contact force (from right to left). BUT NOTE CAREFULLY, that the reduction in contact force takes place *without* any increase in contact resistance. The connection remains sound, but application of a torque wrench at that later time would require further tightening to again achieve the specified torque; and that could go on and on with time. *Do not do any retightening after initial torque setting unless there is some overriding reason.*

In addition to torquing, it is important to follow carefully any instructions on connections that might be supplied by the manufacturer of the connectors or the manufacturer of the equipment that requires connections.

More Than One Conductor Under a Single Lug

Sec. 110-14. **While this *might be* permissible, acceptability of such an application is dependent upon compliance with certain restrictions given in the NEC and the third party testing lab's listing instructions.**

The NEC covers the topic of "dissimilar metals" and connectors (lugs) in Sec. 110-14. Generally speaking, two dissimilar metals, such as copper and aluminum, must not be "intermixed in a terminal or splicing connector"—that is, in physical contact with each other—except where the device is "identified for the purpose and conditions of application." This wording has the effect of *permitting* an application such as that described in the question, *BUT* only when the connector is recognizable as suitable for the purpose, function, use, environment, application, etc.

Typically, this is interpreted to require "listing" or "labeling" by a third party testing lab. And it is the instructions given with the listing or labeling that will indicate what type of markings must be on the lug for a specific application. Remember, because Sec. 110-3(b) requires that all listed and labeled equipment be used in accordance with their listing or labeling instructions, these instructions are actually part of the Code. Failure to use equipment as spelled out in these instructions is technically a Code violation.

The first thing that must be established is: can this lug be used with aluminum and copper conductors in accordance with Sec. 110-14?

The 1988 edition of the U.L. publication, *General Information for Electrical Construction, Hazardous Location, and Electric Heating and Air Conditioning Equipment*, (the so-called "White Book"), under "Wire Connectors and Soldering Lugs" states that a lug is only suitable for use with *copper* conductors unless otherwise marked. If marked "AL," the connector is for use *only* with aluminum conductors. A marking of "AL" and "CU" recognizes the use of *either* copper, *or* aluminum, *or* copper-clad aluminum.

The 1990 edition of the "White Book" now has this information presented in table form as follows:

Marking	For Use With
"AL"	Aluminum
"CC"	Copper-clad Aluminum, only
"AL-CU" or "CU-AL"	Copper, Aluminum, or Copper-clad Aluminum
"CC-CU" or "CU-CC"	Copper or Copper-clad Aluminum
"CU" or "CU-CU"	Copper only
"or equivalent wording"	

The fact that a device is marked "AL-CU" does not in itself recognize the use of copper and aluminum conductors in physical contact with each other. This point is clarified in both editions of the White Book by the statements:

> "Except as otherwise noted on or in the shipping carton, copper, aluminum, and copper-clad aluminum conductors are not to be used in the same connector. A wire connector recognized for securing an aluminum wire in combination with a copper or copper-clad aluminum wire, as noted on or in the shipping carton where physical contact occurs between the wires of different metals, is limited to dry locations and is also marked, i.e. 'AL-CU (dry locations)'."

When a lug is suitable for use with copper and aluminum in physical contact with each other, it is marked "AL-CU (dry locations)" *and* similar marking or wording appears on or in the box. Additionally, if this lug is not in an individual enclosure, the switchboard, panelboard, motor control center, etc. must have some marking, independent of the marking for the lug, indicating its suitability for use with aluminum conductors.

As well as determining if the use of copper and aluminum conductors is acceptable, the suitability of the lug for use with more than one conductor must be established.

The "White Book" states that pressure terminal connectors are intended for only *one* conductor (solid or stranded) unless there is marking on the lug, or, in or on the box in which it came that will specifically indicate how many of what size conductors may be used. Be aware that a marking of, say, #2-1/0 AWG, *does not* mean that more than one No. 2 may be used. This is simply the *range* of wire sizes this lug can accommodate, but only, a *single* conductor of one of those sizes may be used. If it is permissible to use two No. 2's, there must be markings specifically indicating that.

If the lug can be verified as suitable for the use of copper and aluminum in physical contact with each other; with more than a single conductor; and, if not used in an individual enclosure, the switchboard, panelboard, etc. is also independently marked for use with aluminum conductors, then the application would be acceptable.

Another Look at "Clear Work Space"

Sec. 110-16. Rigid interpretation of the rules of NEC Sec. 110-16 resulted in gigantic fines at a large industrial facility.

A large industrial facility that had been operating for many years came under inspection by a governmental agency charged with assuring that such facilities have the safety advantage of compliance with rules of the National Electrical Code (NEC). Experienced and highly competent electrical inspectors from the agency made a detailed and exhaustive inspection of the facility for a number of days and noted many conditions that were not in compliance with rules of the NEC. The biggest surprise of this inspection was the number of citations for violations of NEC Sec. 110-16, which regulates the dimensions of work clearances that are required to afford freedom of movement and safety to any personnel who would be working on the equipment.

An old axiom has it that, "Those who will not learn from history are condemned to repeat its mistakes." Hopefully, everyone will learn the lesson from this report about this large industrial facility that was clobbered with gigantic fines by an electrical inspection authority for failure to satisfy the stringent rules of NEC Sec. 110-16. And in this case, a rigidly enforced deadline for eliminating the violations skyrocketed the cost of retrofit work.

These rules that deal with the code-prescribed "working space" required at electrical equipment are normally thought of strictly as "eyeball" rules—that is, compliance or violation is a matter of placement of equipment with respect to free, clear space around it to permit safe and effective equipment operation, maintenance, and repair. However, as we shall see, knowing when and where to apply which sections of these rules is just as important as understanding the prescribed dimensions. Inasmuch as these rules reach right to the very heart of the overall purpose of the code (i.e. "the practical safeguarding of persons"), careful consideration must be given to certain parts of this section to assure compliance with the "spirit," as well as, the "letter" of the rule.

Each individual who does design or installation is going to have to come to grips with the questions and controversies discussed here and make a decision about how they are going to deal with them. The following is intended to assist in that decision by providing additional information on the real life consequences of these rules.

Working space about electric equipment and working clearances apply to ALL ELECTRIC EQUIPMENT

The most rigid interpretation of the rules in Sec. 110-16 noted was when citations were issued where work benches had been permanently installed in such a manner as to greatly limit access to the receptacles installed beneath these benches. That is, the work benches were considered to be occupying the receptacles' clear work space. As a result all the work benches had to be relocated where they were not within the 30 in. wide, 3 ft. deep, clear work space called for by Sec. 110-16(a).

Was the inspector within his rights? Was he enforcing the Code as written? The answer is yes.

The very first sentence of Sec. 110-16 reads:

> Sufficient access and working space shall be provided and maintained about ALL ELECTRIC EQUIPMENT to permit ready and safe operation and maintenance of such equipment.

This part of Sec. 110-16 has been in the Code for at least 25 years and most electrical designers and contractors feel they understand what it requires. But do they?

The term "equipment" as defined in Article 100 of the NEC specifically includes any and every kind or type of product used "as a part of, or in connection with, an electrical installation." Therefore, the literal wording of this sentence can certainly be taken to extend the requirement for clear work space to junction boxes, outlet boxes, convenience receptacles, wall switches, relays, contactors, etc. As a result, clear work space must also be provided at all such equipment *and* maintained. That is, this work space must always remain clear. It may not be used for storage; and no other equipment, furniture, building columns, other structural elements, etc. may occupy the required dedicated work space.

Some would argue that the rule says that clearances described by the first paragraph in part (a) of Sec. 110-16 are only required where it is "likely" that "examination, adjustment, servicing, or maintenance" might be necessary while the equipment is energized. Although, in the vast majority of cases, a convenience receptacle, for example, could be deenergized prior to any type of repair or maintenance work, in many instances either the nature of the problem or other loads on the circuit may dictate that this repair or maintenance be done while the circuit is energized.

Because the clear work space required by the first sentence and described by part (a) of Sec. 110-16 may be eliminated only where energized examination, adjustment, servicing, and maintenance is not "likely," a look at the definition of this word is in order.

The *American Heritage Dictionary of the English Language* defines "likely" as:

> (1) Having, expressing, or exhibiting an inclination or probability.

This same dictionary defines "inclination" as:

> (1) An attitude or disposition toward something (2) A trend or general tendency toward a particular aspect, condition, or character.

In addition, the definition of "probability" is given as:

> (1) The quality or condition of being probable.

And "probable" is defined as:

> (1) likely to happen or to be true (2) Relatively likely but not certain; plausible.

With these definitions in mind, how is the inspector to determine what is or is not "likely" to occur? That is, how is it possible for the inspector to know the "inclination" of the maintenance individuals with respect to "hot work" or the "probability" of them working on any particular part of the electrical system while energized? Based on everyday real-world experience, we know there can be a variety of extenuating circumstances which might require circuits or equipment to be worked on while energized.

While this may be viewed as extreme, in the inspector's opinion it is not reasonable to assume that these receptacles would only be worked on when deenergized, and, under the prevailing conditions in this shop, maintenance personnel would be exposed to exactly the same types of hazards that the clear work space rules are intended to eliminate.

The lesson here is: be on the safe side of this question. Assume that *all* electric equipment is "likely" to be worked on while energized and provide clear work space, as required in Sec. 110-16(a). This will prevent any "surprises" when an inspection is conducted, as well as limit exposure to legal action in the case of an accident.

In any cases where equipment is installed without diligent attention to required work space, the designer and/or contractor should provide a letter to the owner indicating which equipment has been installed with the work space required by Sec. 110-16(a) because it is assumed that only these components of the electrical system will require "energized" examination, adjustment, servicing, or maintenance. It is important to emphasize that "energized" work on *all* electric equipment that does not comply with the clear work space requirements of Sec. 110-16(a) is unsafe and must be prohibited!

Avoid making any statement that might be construed as indicating that energized work is permissible or safe. Leave that determination to the maintenance personnel. Simply indicate the equipment for which clear work space has been provided.

Maintaining clear work space

In the first paragraph of this section the wording requires that working space be "provided and *maintained.*" Installation of the metal poles within the required clear work space, even though removable, is technically a violation of the requirement for *maintaining* the work space.

This does not, however, eliminate the need for the metal poles, because Sec. 110-17(b) requires some means be provided to protect electrical equipment where it is exposed to physical damage. Certainly, if this equipment is so located as to possibly be struck by a forklift or be exposed to other physical damage, the equipment is exposed to physical damage, and some form of protection must be provided.

Reconciling these two rules has proved troublesome to many individuals for a number of years, especially in industrial facilities where, because of the greater likelihood of physical damage, protection *and* clear work space must be provided.

The ideal solution would be to relocate any obstructions to where they would not be within the work space, thereby complying with the requirement to "maintain" the work space *and* the requirement to provide the necessary "protection from physical damage." (Although, from a design standpoint, it would be better to locate the electrical equipment where it is *not* exposed to physical damage.)

If this is not possible, "special permission" to deviate from a specific Code requirement might be granted by the authority having jurisdiction where "equivalent objectives can be achieved by establishing and maintaining effective safety" as covered by Sec. 90-4.

Be aware that according to the definition of "special permission" as given in Article 100, such permission must be *written* permission. In any instance where special permission is granted by an inspection authority, insist on receiving this permission in writing and keep this document on file in case any questions arise in the future.

Side-to-side clearance

Another point of controversy in Sec. 110-16(a) has to do with the 30 in. width requirement given in the second paragraph. Here, the Code calls for a minimum, 30 in. of elbow room in front of electric equipment. The question is: How should this dimension be measured? Is it permissible to measure this dimension in either of the two ways

Figure 1 It's easy to understand how things like this can happen, especially in an industrial facility where the only constant is change. Nonetheless, an installation like this is in direct violation of the requirements of Sec. 110-16(a) in the NEC. Clear work space must be provided for at least the width of the equipment or 30 in., whichever is greater, and for not less than 3 ft. as measured out from the front of the equipment. And this space must be maintained. Additionally, all doors on such enclosures must be capable of being opened to at least 90°. REMEMBER TO ALWAYS PROVIDE AND MAINTAIN THE REQUIRED CLEAR WORK SPACE.

shown in Fig. 2? Unfortunately, the wording of this Code rule does not address this particular aspect. And experience has shown that some inspectors will accept such an arrangement, while others will not.

But who's right?

On page 34 of the 1989 Technical Committee Report (TCR), it can be seen that proposal 1-110—which is reproduced in its entirety in Fig. 2—addressed this very question. The submitter was asking that Code Making Panel (CMP) 1 clarify the wording for the 30 in. side-to-side clearance minimum. As can be seen in the "PANEL COMMENT," the CMP states it "never intended to prohibit" measuring of the 30 in. from the edge of the equipment, as shown at the right of the drawing submitted with the proposal. This is difficult to reconcile with the concept of "elbow room" for side-to-side clearance from live or grounded equipment as discussed in the original proposal for side-to-side clearance, which appeared on page 11 of the "PREPRINT" OF THE PROPOSED AMENDMENTS FOR THE 1971 NEC.

It is clear from Proposal No. 38, as it appeared in the '71 Code "PREPRINT," that the submitter was concerned with the fact that

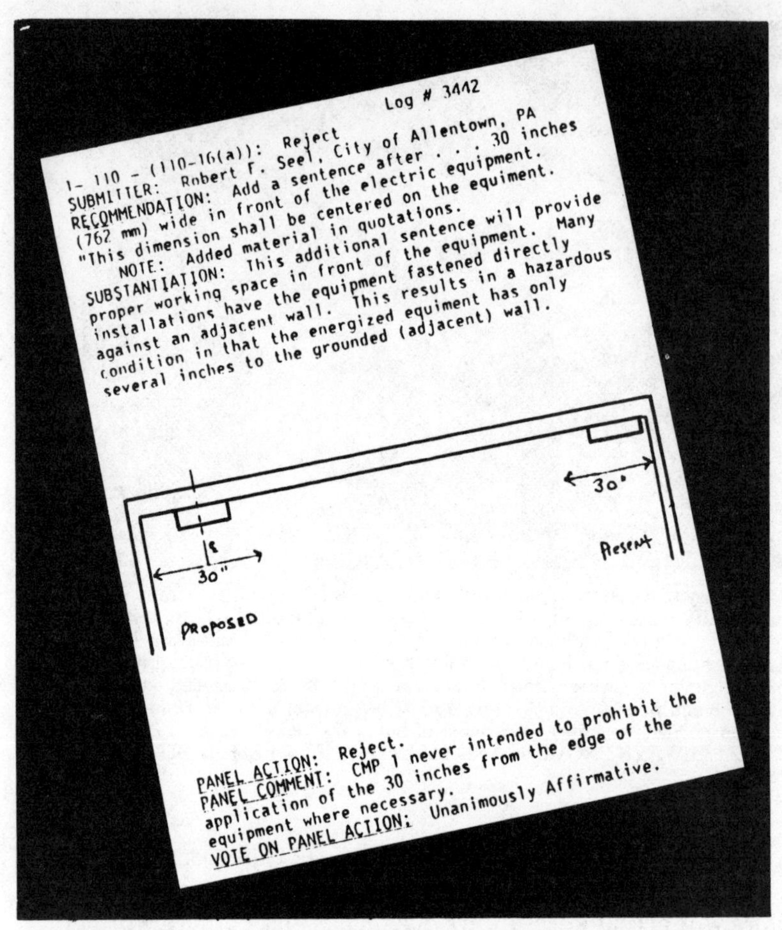

Log # 3442

1- 110 - (110-16(a)): Reject
SUBMITTER: Robert F. Seel, City of Allentown, PA
RECOMMENDATION: Add a sentence after . . . . 30 inches (762 mm) wide in front of the electric equipment. "This dimension shall be centered on the equiment.
NOTE: Added material in quotations.
SUBSTANTIATION: This additional sentence will provide proper working space in front of the equipment. Many installations have the equipment fastened directly against an adjacent wall. This results in a hazardous condition in that the energized equiment has only several inches to the grounded (adjacent) wall.

PANEL ACTION: Reject.
PANEL COMMENT: CMP 1 never intended to prohibit the application of the 30 inches from the edge of the equipment where necessary.
VOTE ON PANEL ACTION: Unanimously Affirmative.

Figure 2 This proposal and drawing were submitted during the Code-cycle for the 1990 edition of the NEC. The "PANEL COMMENT" makes clear that the panel's intent is to permit applications such as that shown at right in the drawing. While such an application may be permitted by the Code, each individual designer and installer must carefully consider the potential consequences of utilizing such permission.

while Sec. 110-16 and Table 110-16(a) required specific clearances from grounded surfaces or energized parts in the direction of access—that is, at a right angle to the front of the equipment—to minimize the possibility of shock or electrocution during maintenance of energized equipment, there were no requirements for *side-to-side* clearances from grounded surfaces and energized parts. Why require clearance from such surfaces and parts in one direction and then provide for no clearance in the other direction? In an attempt to rectify this problem, the following wording was proposed:

"There also shall be a minimum clear working space of 1 1/2 ft. (measured at right angles to the direction of access) to any grounded surface or other live parts. Concrete, brick, or tile walls shall be considered as grounded."

The supporting comment went on to point out that 1 1/2 ft. "dimension approximates the distance from an extended elbow to the center of a person's body, which might be centered in front of the point of work."

This proposal, which was first submitted during the '68 Code-cycle, was held over to the '71 Code-cycle to allow comment on the specific dimension. Proposal No. 43 was submitted to address the concern for the specific dimension and it called for the work space to be "at least 30 in. wide in front of the equipment."

CMP 1 then combined these two proposals and came up with the following wording, which appeared for the first time in the Sec. 110-16(a) of the 1971 NEC.

"In addition to the dimension shown in Table 110-16(a) the work space shall be at least 30 in. wide in front of the electrical equipment. Distances are to be measured from the live parts if exposed or from the enclosure front or opening when such are enclosed."

Refer to Fig. 3. If the 30 in. wide *clearance* is measured from the center of the panelboard, then maximum separation from grounded surfaces and other live parts *on both sides* of the equipment would be provided. This is more consistent with the general concept of "work

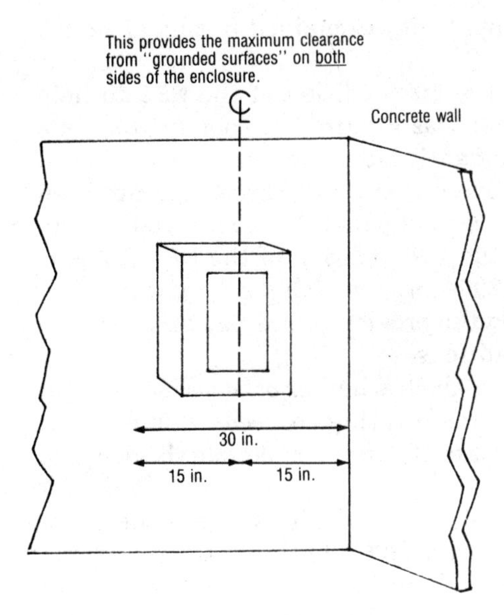

Figure 3 This application is more consistent with the original concern behind the 30 in. wide clear work space requirement. If the panel were closer than 15 in. to the concrete wall, accidental contact with this "grounded surface" would expose the mechanic to increased shock or electrocution hazards during maintenance of the energized panelboard. Although CMP 1 has said that installation of such a panelboard with its side flush against the wall is *not* intended to be prohibited, real safety can only be achieved by providing "clearance" from exposed live parts and grounded surfaces, which is the basic thrust of the requirements in Sec. 110-16(a) and Table 110-16(a).

space" and the logical understanding of "elbow room," which would call for clear space on *both* sides of the equipment. In addition to permitting the use of both hands and providing for less physically-restricted movement, a mechanic would be less likely to come in contact with the grounded surface—the concrete wall—and the possibility of shock or electrocution while working on the energized equipment would be reduced, which is the basic intent of Sec. 110-16(a).

Although it is the intent of the CMP to permit installations such as that shown at right in the drawing of Fig. 2, remember that the NEC is a *minimum* standard. It is expected that in addition to complying with the bare minimum requirements of the NEC, we must exercise good judgment, recognize those instances where the Code might be construed to be deficient—as could be the case with this rule—and go beyond that which would be permitted by the Code.

Even if the CMP is no longer concerned with clearance from grounded surfaces or exposed live parts on either side of the equipment, by measuring 15 in. in each direction from the centerline of the equipment and providing "elbow room"—that is, clear space on both sides of the equipment—the installation cannot be called into question by anyone. Such an approach would also show "good faith" on the part of the designer or installer, which may be the difference between an acquittal or guilty verdict should the designer or contractor ever be involved in a legal action.

Access and entrance to working space

Does Sec. 110-16(c) cover "Line-ups" with Combined Rating of 1200A or More?

Sec. 110-16(c) was revised in the 1990 edition of the NEC to help clarify the type of equipment that was required to comply with this part of Sec. 110-16. It is now abundantly clear that this rule applies to any equipment that: (1) contain overcurrent, switching, or control devices [not just "switchboards and control panels" which were the only type of equipment identified in the 1987 edition of the NEC], (2) are over 6 ft wide, and (3) rated at 1200A or more. This rule now covers a variety of applications that, given the previous wording, were not covered under this section in past editions.

However, there is still a question about whether or not a "line-up" of equipment, whose combined individual ratings exceeds 1200A, is required to comply with the "two points of entry" or "double the depth of the work space" rules.

Consider an application such as shown in Fig. 4. Here there are three, 400A panelboards tapped from a 1200A feeder in the trough be-

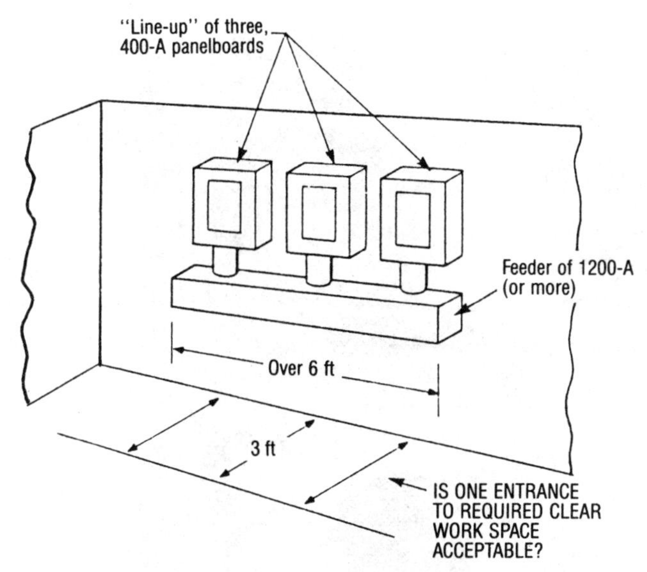

Figure 4 Is this application covered by the requirements of Sec. 110-16(c)? Although not clearly spelled out in this Code rule, providing for two entrances or doubling the depth of work space in such an application certainly *cannot* be considered a violation, and it may save someone's life!

low. Is this application required to comply with Sec. 110-16(c)? The wording of the Code rule does not clearly address this matter. Should this equipment be provided with two entrances or twice the work-space depth required? Without a doubt, the smart money says, yes!

This rule is essentially aimed at assuring more than one means of access to the work space at electrical equipment when the amount of current available is 1200A or more. It recognizes that there is a greater hazard because of the relatively large amount of energy that could be delivered into a fault at that point in the system. In fact, this rule was enacted as the result of a number of fatalities where maintenance personnel were trapped in the work space when an arcing burn down erupted in their path of exit. (See Fig. 5.) Had there been another exit route, these individuals could have escaped and almost definitely would not have been killed. Because the 1200A feeder supplying the panelboards in Fig. 4 could deliver the same amount of energy into a fault as 1200A rated busbar within a switchboard, it would seem unwise to suggest that the three 400A panelboard line-up does not require the same amount of work space *and* number of entrances as a 1200A switchboard would. Be on the safe side. Provide access to the working space as called for in Sec. 110-16(c) at any location where

Figure 5 This is exactly the type of installation that has caused the deaths of a number of individuals over the years. Because there is only one way to get into or, more importantly, out of the defined work space, if an arcing burndown occurs in the section of the switchboard between the mechanic and the work space entrance, the mechanic would be trapped. As a result, Sec. 110-16(c) requires either two entrances or doubling the depth of work space required by Table 110-16(a).

the equipment is over 6 ft. wide, fed from a single source (busbar, feeder, branch-circuit, or feeder tap) rated at 1200A or more, and contains overcurrent, switching, or control devices.

Rules on "headroom" and "illumination" should be extrapolated to assure genuine safety

Two other parts of Sec. 110-16 also require some serious consideration and analysis prior to application. In Secs. 110-16(e) and (f), the Code calls for "illumination" and "headroom" for the working spaces about "service equipment, switchboards, panelboards, and motor control centers." These rules place additional requirements on the work space about these types of equipment.

As given in Part (e), the Code requires that some type of lighting be provided to illuminate the clear work space at *indoor* service equipment, switchboards, panelboards, and motor control centers. And Part (f) calls for the work space area to be clear and free from obstructions to a height of at least 6 ft 3 in. (See Fig. 6). These two "make-sense" propositions are intended to further enhance personnel safety.

By calling for lighting of this equipment, maintenance personnel

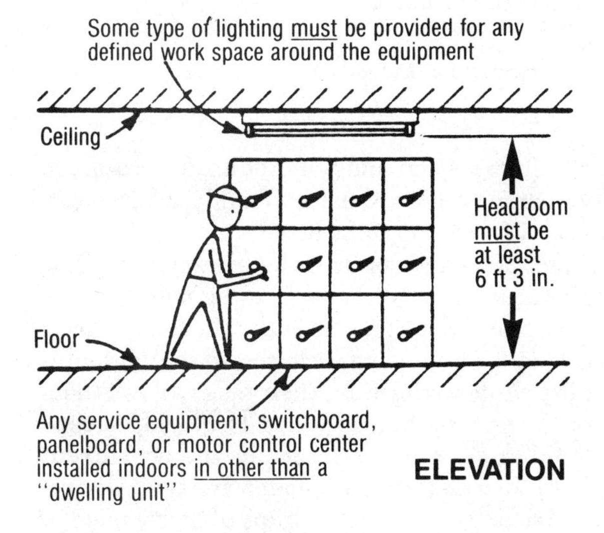

Figure 6 The basic rule of Secs. 110-16(e) and (f). While the wording of the rule indicates it applies only to "service equipment, switchboards, panelboards, and motor control centers," adequate headroom and lighting should also be provided at individual CBs and fusible switches, especially those rated 200A or more, to assure maximum safety during maintenance and repair.

Figure 7 If the disconnect mounted above the transformer in this photograph were used as a service disconnect, then the transformer would be prohibited by Sec. 110-16(f) from being installed beneath the individual disconnect switch. Even though this disconnect switch is *not* used as service equipment, the same hazards would exist and the requirements of Sec. 110-16(f) and (e) should be extrapolated and voluntarily applied.

will always have adequate illumination and will not have to resort to the use of a drop light, which may or may not be readily available, a hand-held flashlight, or, even worse, no light at all.

The headroom requirement of Part (f) assures that personnel will be able to stand in front of the equipment while performing maintenance or repairs without having to crouch, or bend, or drape themselves over another piece of equipment. Providing enough clearance so that maintenance personnel can stand while performing their tasks allows them to be in a better position to work comfortably and to quickly escape the area if a problem should arise. And, inasmuch as the noncurrent carrying metal parts of all other electric equipment are typically required to be grounded by Article 250, having to drape over one piece of equipment to work on another piece of equipment presents the same type of hazard that is addressed by Condition 2 of Table 110-16(a), and, correctly, should be avoided.

Because these rules also relate to the very important consideration of providing for personnel safety, the question is: Should the requirements given here for service equipment, switchboards, panelboards, and motor control centers be voluntarily applied to individually enclosed switches and circuit breakers as well? The only rational answer is yes, especially for those switches and CBs rated 200A or more.

Refer to Fig. 7. In this application, the rules of Secs. 110-16(e) and (f) would very definitely apply if the individually enclosed fusible switch above the transformer were a service disconnect. And the transformer would be prohibited by Sec. 110-16(f) from occupying the 6 1/4-ft high work space—which must extend from the floor to that height—dedicated to the fusible switch. If headroom and illumination is required to assure safety during maintenance when the fusible switch is used as a service disconnect, then logic dictates that these rules should also apply when the switch is used as other than a service disconnect.

To do otherwise is really not consistent with personnel safety and would be very difficult to defend in court. If someone were injured or killed because they were in contact with the grounded transformer case while working on the switch shown in Fig. 7, the contractor and/or designer could be criticized if the only reason for failing to provide the headroom called for in Sec. 110-16(f) is "the Code didn't require it."

The same argument applies to the Exception given after each of these rules as well. Residential services need light and clear work space, too! Simply because the Code would permit otherwise ought not be sufficient reason to eliminate these requirements. To protect yourself, *do not* apply the permission given in the Exception, unless absolutely, positively necessary.

Remember, it is expected that in addition to complying with the bare minimum requirements of the NEC, we must exercise good judgment, recognize those instances where the Code can be construed to be deficient, and, in those cases, provide more than just what is called for by the Code.

Protect yourself and those who will service and maintain the installed equipment. Always provide clear work space at ALL electric equipment.

Figure 8 Work clearance in front of this thermostat and other electrical enclosures seems to be free and safe. But the conduits and electrical enclosures are recessed in the web space of a vertical, structural "I"-beam with the flange of the beam protruding forward further than the depth of the boxes. With that arrangement, the forward edge of each beam flange (left and right) protrudes into the space that is required to be clear—that is, into the space 30 in. wide and flush with the covers of the enclosures. Although the flange edges protrude only slightly into that space, and do not present any obstacle to side movement of a worker's elbows, the condition is literally in violation of the letter of the rules of Sec. 110-16(a)—both of which combine to require a clear work space that is 3 ft. deep (front to back) and no less than 30 in. wide in front of the equipment. The extended flanges of the beam are literally excluded from that space. As with so many other NEC rules that specify dimensions and numerical quantities; there is no permission to disregard "small" variations from the dictated numbers, even the few inches involved here. The best practice is to adhere rigidly to all Code maximum and minimum dimensions.

Locked Enclosures Are "OK"

Sec. 220-24. Although the basic rule of Sec. 240-24 calls for fuses and circuit breakers that provide Code-specified overcurrent protection to be "readily accessible," when judging whether electrical equipment is "readily accessible" it is first necessary to determine if the equipment is "accessible" in accordance with the NEC's meaning of that word.

Note that in Article 100 the definition of "accessible" (as applied to equipment) does not prohibit a locked enclosure or equipment room door, but rather, requires that locked doors, where used, not "guard" against access. That is, the keys for such locks must be in the possession of or available to those requiring access so that the locked room or enclosure door does not prevent access. Moreover, by referring to "locked doors," the definition actually presumes the existence and, therefore, the acceptability of "locked doors" in electrical systems. In addition, as given in the definition for "readily accessible," any equipment that has to be "readily accessible," only must be so for "those to whom ready access is requisite." This clearly recognizes making equipment inaccessible to other than authorized personnel, such as by providing a lock on the door and having the key possessed by or available to those who require access. When keys are in the possession of or available to those requiring access, locking of a panelboard or installation of a panelboard in a locked room is not prohibited by Sec. 240-24.

Does the Code Require Limiting of Voltage Drop?

Secs. 215-2 and 210-19(a). No, the NEC does not *require* the upsizing of the neutral conductor when the phase conductor is up-sized to prevent excessive voltage drop, but this is commonly done. While it is common practice to "automatically" increase the size of the neutral conductor to that of the phase conductor to limit voltage drop on long runs, it *is* possible to reduce the voltage drop to an acceptable level *without* increasing the size of the neutral. Refer to Figure 1.

The NEC does not even *require* up-sizing of the *phase* conductors for voltage drop. As covered in the Fine Print Notes (FPN) following Secs. 210-19(a) and 215-2, the Code only *recommends* that voltage drop be limited to a 3% maximum on any branch circuit and a 5% maximum overall (i.e., feeder and branch circuit, combined).

It is worth noting, however, that the NEC in Sec. 250-95 *does* require the size of the equipment grounding conductor to be "adjusted proportionately according to circular mil area" when the circuit conductor(s) are up-sized for voltage drop. This is intended to assure that the equipment grounding conductor is no less than approximately 12.5% of the cross-section area and 20% of the conductivity of the phase conductor. These parameters, which are the basis for Table 250-95, have been empirically determined as acceptable "minimums" to provide the

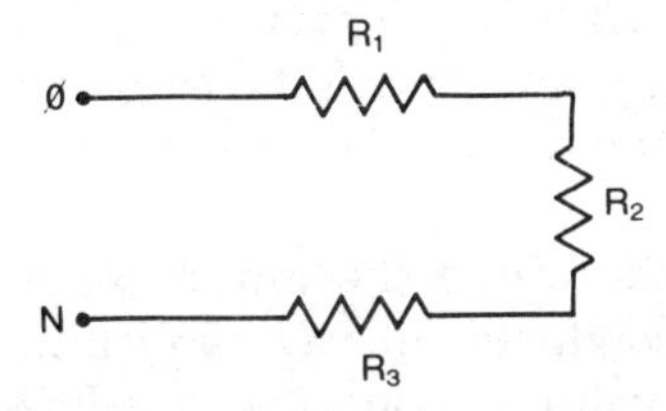

Where:

R_1 = Resistance of Ø conductor
R_2 = Resistance of the load
R_3 = Resistance of the neutral conductor

Because:

E_{VDT} (total voltage drop) = $(I \times R_1) + (I \times R_2) + (I \times R_3)$
If R_1 is reduced by up-sizing of the Ø conductor,
E_{VDT} is also reduced.

Figure 1

Code-required low-impedance fault return-path necessary to facilitate operation of the circuit overcurrent protective device. Therefore, when the designer or contractor determines a circuit to be so long as to introduce excessive impedance and require an increase in size for the phase conductor, the size of the equipment grounding conductor required by Sec. 250-95 must also be increased "proportionately." Increasing the size of the equipment grounding conductor will prevent an excessive impedance increase of the conductor due to the long run. This is accomplished as shown in the following example.

Given The phase conductor on a 20A branch circuit is up-sized from No. 12 AWG, solid copper to No. 10 AWG, solid copper.

Find The minimum acceptable size of equipment grounding conductor.

1. From Table 8 in Chapter 9, determine the cross-sectional area in circular mils: No. 10 AWG, solid copper conductor = 10380 cmils; No. 12 AWG, solid copper conductor is 6530 cmils.
2. Divide the cross-sectional area of the up-sized phase conductor (No. 10 solid copper) by that of the required size of phase conductor (No. 12 solid copper) to determine the proportional factor.

$$10380/6530 = 1.5895$$

3. From Table 250-95, determine the size of equipment grounding conductor required for a 20A branch circuit.

No. 12 AWG, copper

4. Multiply the cross-sectional area of the equipment grounding conductor called for in Table 250-95 by the proportional factor determined in Step 2.

$$6530 \text{ cmils} \times 1.5895 = 10379.435 \text{ cmils}$$

Using Table 8 in Chapter 9 we find that a conductor with at least 10379.435 cmils is a No. 10 AWG. Therefore, the minimum size equipment grounding conductor must be a No. 10 AWG, copper conductor.

We do not want to talk about how many angels can dance on the head of a pin; the above simply conveys the concept behind the word "proportionately."

This same procedure can be used to determine the minimum required size for an equipment grounding conductor in any application where circuit conductors are up-sized for voltage drop considerations.

Grounding Metal Conduits on Service Conductors Run Down Pole

Secs. 250-32, 250-51, 250-61(a), 250-91(c), and 300-5(d). Much controversy is often generated by the subject of grounding metal conduit that is used as a protective sleeve on service conductors brought down a pole for an underground lateral supply. Article 250 has a number of very clear and widely understood rules that apply to this consideration. Figure 1 shows a typical application of the equipment under discussion here.

In the common layout for an underground service from a pole line to a building, the conductors are run down the pole in nonmetallic raceway to a point at least 8 feet above the ground—at which point metal

Figure 1 Steel conduit lengths (arrow) enclose service entrance conductors to building at left, with the steel conduits extending from below grade to 8 feet above grade where they connect to plastic conduits that enclose the conductors for the remainder of the run up the pole to connect to the supply transformers. Underground from the pole to the building, the service lateral conductors are run in PVC conduit. It is a Code requirement and an important safety necessity that the steel conduit lengths exposed to contact by persons be properly grounded by connection to the grounded neutral conductor. Such connection will assure that any accidental contact between one of the energized service conductors and the metal conduit will be burned clear and will not leave a hazardous shock voltage on the metal conduit.

conduit is used for the run down and into the earth. Section 300-5(d) requires that the conductors be mechanically protected from below ground up to at least 8 feet aboveground by rigid metal conduit, intermediate metal conduit (IMC), or schedule 80 rigid nonmetallic conduit. The underground run of conductors is then made directly buried or in nonmetallic conduit. In those cases where rigid metal conduit or IMC is used to protect for the first 8 feet up the pole, the questions arise: (1) Must the metal conduit run be grounded? and, if so, (2) How must it be grounded?

NEC Sec. 250-32 says that "Metal enclosures for service conductors...shall be grounded," and that rule applies to service "raceways." As a result the metal conduit that encloses service conductors on a pole must be grounded. There are no exceptions given in Sec. 250-32.

The basic elements of this problem must be clearly understood to properly satisfy the safety intents of the Code rules that apply:

1. A metal raceway enclosing service conductors run down a pole or on the outside of a building is always exposed to the possibility that one of the enclosed energized conductors might sustain an insulation failure that places the metal of the conductor in contact with the metal of the conduit.

2. When such contact between conductor and conduit takes place a voltage (phase-to-neutral voltage) will be placed on the metal conduit. If there is not a low-impedance equipment grounding connection between the conduit and the neutral of the supply circuit, the fault connection will remain intact and the voltage on the conduit will pose a shock or electrocution hazard to persons who might touch the conduit and have another part of their body in contact with earth or some other metal in contact with earth.

3. The danger of that situation is not at all reduced by connecting the metal of the conduit to a driven ground rod at the pole. Because service conductors do not usually have overcurrent protection at their supply ends, there will be no clearing of the fault by opening of a fuse or circuit breaker. And even if there were overcurrent protection in the circuit, the impedance of the earth would keep fault current far too low (milliamps or microamps) to operate any fuse or CB. (See the last sentence of Sec. 250-51, which notes the inadequacy of earth current flow.)

4. The only way to clear the dangerous potential from the service conduit is to provide an equipment grounding connection between the conduit and the neutral conductor. If the metal conduit is connected to the circuit neutral, any phase conductor coming in con-

tact with the conduit will produce a phase-to-neutral fault that will carry enough current to burn open the point of fault and thereby remove the dangerous voltage (and shock hazard).

5. This whole scenario emphasizes the need to provide effective return from metal enclosures to the grounded neutral conductor of a grounded electrical system. It also pinpoints the fallacy that mere connection to earth (such as to a grounding electrode) will provide for fault clearing and assure safety to persons. The last sentence of Sec. 250-51 and the last phrase of Sec. 250-91(c) prohibit use of the earth as a path of fault current flow—that is, the earth must not be used as the "sole equipment grounding conductor."

Because NE Code Sec. 250-61(a) permits use of "a grounded circuit conductor" (that is, the grounded neutral conductor of a circuit) to ground service conduits, such connection is the most effective way to satisfy Sec. 250-32, requiring grounding of metal service conduits. But the question immediately arises—"How can that be accomplished?"

The first and most obvious method that could be used to bond metal service conduit to the neutral would be a bonding jumper from a lug on a bonding bushing on the top end of the service conduit length—with the jumper connecting from the bushing lug to a tap connector on the grounded neutral (Figure 2). With such a connection, the metal conduit would be grounded, as required by Sec. 250-32, without dependence on earth as an equipment grounding conductor. Or, an equipment grounding conductor could be run from the bonding bushing up to a neutral connection at the top of the pole at the supplying transformer.

Figure 2 Metal service conduits (containing service entrance conductors running from the transformer secondary terminals to the building) are properly grounded here (left arrow) by an insulated copper conductor connected to lugs on bonding bushings on the conduit ends. The equipment grounding (or bonding) conductor connects down and under the crushed stones to the grounded neutral terminal bus (right arrow), which is also bonded to the angle-iron mounting frame. Although underground raceways are always assumed by the NEC to be filled with water, the installer did pack the upper ends of the conduits with duct-seal compound to satisfy the literal rule of Sec. 230-53, which requires that service raceways "exposed to the weather"..."shall be raintight."

One Steel Column OK as Grounding Electrode

Sec. 250-81. When structural building steel is used as a grounding electrode, connection to only *one* steel column will satisfy the letter and intent of this Code rule.

According to the basic rule of this section, all or any of the grounding electrodes described in parts **(a)**, **(b)**, **(c)**, and **(d)** that are available on a premises must be bonded together to form a "grounding electrode system"—which is then connected by a grounding electrode conductor to the neutral and/or ground bus of the premises electrical system. Although the wording of all parts of Sec. 250-81 is clear and straightforward, much controversy often arises over the rule of part **(b)**, which requires that the "metal frame of the building, *where effectively grounded,* must be used as one of the electrodes."

Where a building has a structural metal frame of vertical and horizontal I-beams (or H-beams), all the metal lengths are welded or riveted together to form a single interconnected assembly. And at the concrete footing or foundation of the building, each of the vertical beams is connected into the concrete by means of a J-bar that bolts the bedding plate of the beam to the concrete. Then because the entire mass of concrete and its reinforcing are set in the earth and form a huge grounding electrode, all the vertical structural steel columns become part of that electrode as a result of their very low-impedance bonding to the concrete. In fact, the multiple connections between the vertical columns and the concrete effectively make the entire assembly of the building structural steel part of the "grounding electrode system."

Because all the steel columns are bonded to the concrete mass, all the columns are therefore bonded to each other, and there is no need (and also no Code requirement) to connect bonding jumpers from column to column. When the grounding electrode conductor from the service equipment connects to one of the steel columns or to one of the other electrodes that are required by Sec. 250-81 to be bonded to one of the columns, then the grounding electrode conductor, in fact, is connected to all the steel columns and to the rest of the building steel structure.

As a result of the above-described condition, connection of a service grounding electrode conductor or bonding jumper from one of the other electrodes may be made at any one point on any column. Or, a

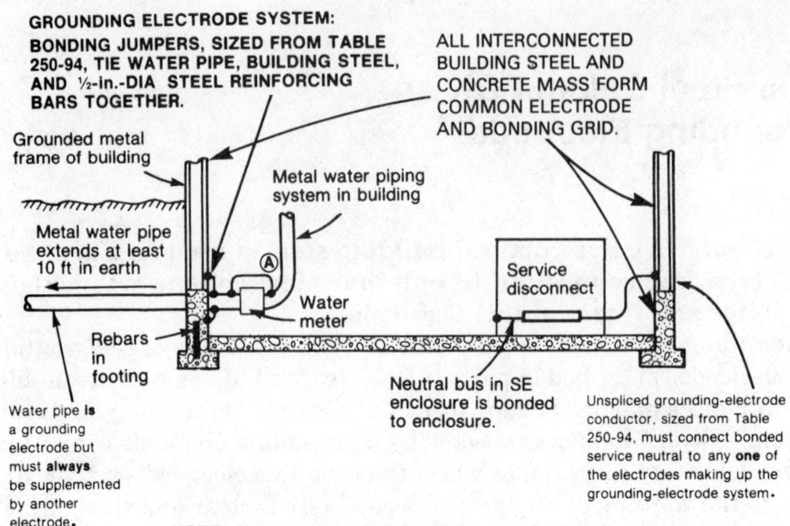

THE BONDED SERVICE NEUTRAL HERE MAY BE CONNECTED BY GROUNDING ELECTRODE CONDUCTOR TO EITHER THE BUILDING COLUMN AT RIGHT OR AT LEFT — OR ANY OTHER COLUMN THAT IS PART OF THE BUILDING STRUCTURE.

Figure 1

grounding electrode conductor might be connected to one column close to the service equipment and, say, a water-pipe electrode that is a distance away from the service equipment may be bonded to another column near the water pipe—taking advantage of the fact that the building steel forms a "common bonding grid." (See Figure 1.)

Sizing Equipment Grounding Conductors Run in Flex or Liquidtight

Secs. 250-95, 350-5, and 351-9. Sizing an equipment grounding conductor for primary and secondary circuits of a transformer when the circuit conductors are run in flexible metal conduit or in liquidtight flexible metal conduit.

One of the most common applications of flexible metal conduit and liquidtight metal conduit is to enclose the short lengths of circuit conductors connected to the primary and secondary terminals of low-voltage (up to 600 V), dry-type transformers so commonly used in modern buildings to step 480 V down to 120 V. Typically, such transformers are either single-phase units stepping 480-V, single-phase down to 120/240-V, 3-wire, single-phase, or they are 3-phase units stepping 480-V delta down to 208Y/120 V, 3-phase, 4-wire. In all such cases a number of controversies have been generated by Code rules applying to flexible metal raceways. Problems arise with respect to the need for equipment grounding conductors, sizing of such conductors, and the need for clamping or securing the flexible metal raceway in place.

Flexible metal conduit (standard or liquidtight) is used to make such transformer connections because such usually short connections are much more easily made with flex than with any rigid conduit, and the flexible connection provides important vibration isolation to reduce the noise (the hum) from the transformer.

Figure 1 shows a typical application of flex for transformer connections. In any particular case, the specific components may vary in type and size, but the general layout will be as shown, and the various NE Code considerations will be the same.

In the layout shown in Fig. 1, a short length of flexible metal conduit is used to enclose the primary feeder conductors to the transformer. The 30-kVA transformer has a rated primary current of 36 A (30,000 VA – 480 V × 1.732) and must have a primary protective device rated not over 125% × 36 A—or 45 A. That calls for a 45-A, 3-pole CB or a switch with 45-A fuses [Sec. 450-3(b) (1)]. Then the circuit conductors for the primary 480-V feed to the transformer must be sized to be properly protected by the 45-A protection in accordance with the conductor ampacity—which means that the primary circuit conductors must have an ampacity of 45 A.

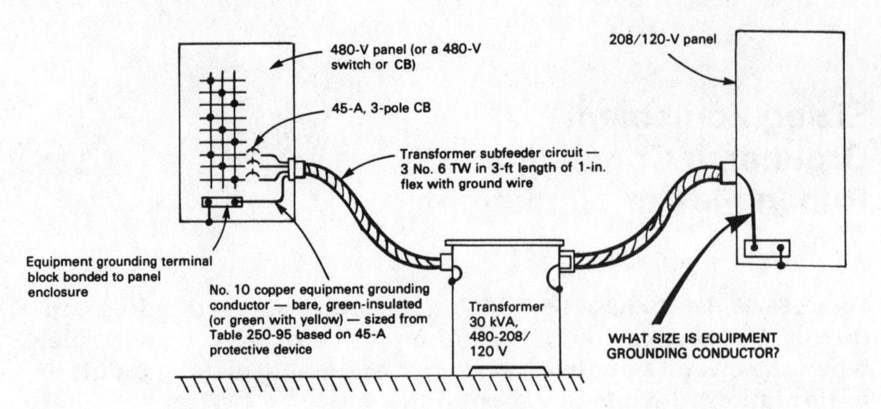

Figure 1 Equipment grounding conductor is always required for unclamped flex connections.

The circuit conductors to the transformer primary must be at least **three No. 6 TW (or THW, THHN, RHH, or XHHW) copper** in flex. The flex must be 1-in. size for TW, THW, and RHH or ¾-in. size for THHN or XHHW. As required by UL rules, any circuit breaker rated up to 125 A is listed for use with conductors operating at not over the 60°C ampacity shown in NE Code Table 310-16 for the particular size of conductor. (If the 45-A CB in the 480-V panel is UL-listed as suitable for use with 75°C wires—and is so marked—it would be acceptable to use No. 8 THW, THHN, RHH, or XHHW conductors for the primary feed to the transformer. Those wires have an ampacity of 50 or 55 A in Table 310-16 and would be well protected by the 45-A CB.)

The length of flexible metal conduit used to carry the No. 6, 3-phase circuit from the 480-V panel to the transformer, which is mounted close to the panel, must have an equipment grounding conductor run in the flex, regardless of the length of the flex, because the conductors contained in the flex are protected by overcurrent devices rated over 20 A. This is required by Secs. 350-5 and 250-91(b). A length of flex up to 6-ft long may be used without a separate equipment grounding conductor—only if the conductors in the flex are protected at more than 20 A, as stated in Exception No. 1, Sec. 350-5.

The transformer circuit in Fig. 1 is protected at 45 A; therefore a circuit in flex would require an equipment grounding conductor sized from Table 250-95 of Sec. 250-95. The table shows that for a circuit protected by an overcurrent device of 30-, 40-, or 60-A rating, a No. 10 copper or No. 8 aluminum grounding conductor must be used. The 45-A protection, therefore, calls for that size of equipment grounding conductor, which must be run from an equipment grounding terminal in the primary-side enclosure (panelboard, CB, or switch) to an equipment grounding terminal in the transformer enclosure. The purpose of that grounding conductor is to conductively connect the metal enclosures that are connected by the flex because the flex itself is a poor grounding path.

Big problem!

As shown in Fig. 2, it is usual practice to also use flexible metal conduit to carry the secondary 208/120-V transformer feeder to the panelboard (or fused switch, circuit breakers, motor control center, or whatever the transformer is feeding). But, on the transformer secondary, the feeder conductors to the panel come right off the terminal lugs (marked on the transformer as X_0, X_1, X_2, and X_3). There is no overcurrent protection for those conductors at their point of origin at the transformer. Because these secondary conductors are not "protected by overcurrent devices rated 20 amps or less," the rules of Sec. 250-91(b) and Sec. 350-5 would again require that an equipment grounding conductor be run with the circuit conductor to provide an equipment ground path from the transformer enclosure to the metal panel enclosure.

But, how do we size the equipment grounding conductor on the transformer secondary?

In all such transformer secondary hookups of flexible metal conduit, the equipment grounding conductor must be sized in accordance with the rule of Sec. 250-95, which calls for the grounding conductor size to be selected from Table 250-95, based on the rating of the overcurrent device protecting the hot circuit conductors in the flex. But, as noted above, there is no overcurrent protection at all for the secondary feeder conductors from the transformer to the panel. The absence of that protection is permitted by Exception No. 2 of Sec. 240-21, covering use of an unprotected 10-ft tap. (Such practice is also recognized by Exceptions No. 3, No. 8, and No. 11 of Sec. 240-21.) But, the absence of overcurrent protection for that circuit makes it impossible to size the equipment grounding conductor for the secondary flex connection from Table 250-95.

The only solution for this problem is to base the size of the equipment grounding conductor on the rating of an overcurrent device that would properly protect the secondary feeder phase conductors at their ampacity. In effect, we determine what rating of overcurrent protective device would protect the secondary phase legs and then use that value of protection in Table 250-95 to determine the minimum required size of equipment grounding conductor.

In the diagram of Fig. 1, the 30-kVA transformer has a rated secondary current of 83 A. Using conductors at their 60°C ampacity (the TW amp ratings), the smallest size of conductor that would handle the full-load output of 83 A would be No. 3 copper with TW insulation (or THW, THHN, RHH, or XHHW insulation on a No. 3 copper conduc-

Figure 2 Short flex lengths are used here at the connections for the primary and secondary circuit connections to a transformer. Because flex in this application does not satisfy Exception No. 1 of Sec. 250-91(b) or Exception No. 1 of Sec. 350-5, an equipment grounding conductor must be used to assure a low-impedance ground-fault current path through each flex length. If an equipment grounding conductor is run within the primary and secondary raceway connections, a grounding conductor must be run from the panel where the primary circuit originates, within the rigid metal raceway (at top) and through the flex to the transformer equipment grounding terminal within the transformer enclosure. That conductor would have to be run the entire length even though the rigid metal conduit is itself a suitable equipment grounding conductor. But the long length of equipment grounding conductor is required because the primary flex length (in right foreground) interrupts the continuity of the rigid metal conduit to the transformer case. And the same reasons would apply to require a secondary-side equipment grounding conductor from the transformer case to the secondary panel (lower left). The grounding conductor in the primary circuit would be sized from Table 250-95, based on the rating of the CB or fuses protecting the hot conductor of the primary feeder. But the secondary circuit grounding conductor would have to be sized as described in the accompanying text—by simply using the rating of the protective device that would properly protect the secondary circuit hot legs.

Note: Section 250-79(d) and (e) would permit use of an equipment bonding jumper to be installed around each short length of flex by connecting a jumper from a lug on the flex connection at one end of the flex to a lug on the flex connector at the other end of the flex. But such flex connectors with an external conductor terminal would have to be UL-listed fittings, and the length of such an external bonding jumper (which functions exactly the same as an equipment grounding conductor) must not exceed 6 ft and must be tied to the flex. The size of such external bonding jumpers would have to be the same as required for an equipment grounding conductor run inside the raceways. The external bonding jumper completes the ground path continuity from the rigid metal raceway to the transformer enclosure.

tor). To protect No. 3 TW in accordance with its 85-A ampacity, as required by Sec. 240-3, a fuse or CB pole rated at no more than 90 A would be required. (In Sec. 240-6, 90 A is the maximum standard protection rating above the 85-A conductor ampacity—which is permitted by Exception No. 1 of Sec. 240-3.)

Referring to Table 250-95, an overcurrent protective device rated over 60 A but not over 100 A would call for an equipment grounding conductor not smaller than No. 8 copper or No. 6 aluminum.

Therefore use of a No. 8 copper equipment grounding conductor in the secondary flex run would satisfy the concept of Sec. 250-95 on sizing of equipment grounding conductors. And No. 8 copper would be a minimum no matter what insulation might be used—TW, THW, THHN, etc.

Another problem!

In the very widely used type of installation shown in Fig. 1, it is the usual practice to leave the flex runs *without* any clamp or other means of securing the flex to the building surfaces. Section 350-4 basically requires that flex be secured "by an approved means at intervals not exceeding 4½ feet and within 12 inches on each side of every ...fitting." Exception No. 2 relaxes that rule "where flexibility is necessary," by permitting lengths of not over 3 ft to be used without clamping. And it is generally accepted that transformer connections require flexibility for vibration isolation.

But note that the maximum length that may be used without clamping is *3 ft.* Use of an unclamped length of flex longer than 3 ft is clearly a violation of this Code rule. In the example here, neither the primary flex run nor the secondary flex run may be longer than 3 ft if left unclamped. However, it would satisfy the Code rule if a 4-, 5-, or 6-ft length of flex was used with a clamp at the midpoint of the run so that the unclamped length at each terminal was not more than 3 ft long.

Note: Exception No. 3 of Sec. 350-4 permits up to a 6-ft length of flex without need for any clamps, *but* that use is limited to connection of lighting fixtures when installed as permitted by Sec. 410-67(c). The 6-ft unclamped length does not apply to transformers, motors, or any other equipment.

NEC rules on use of liquidtight flexible metal conduit are generally similar to the above rules on standard flexible metal conduit, with the following qualifications:

1. As a suitable grounding means, liquidtight flex is regulated by Exception No. 2 to Sec. 250-91(b) and by Sec. 351-9. Liquidtight flex is

permitted as an equipment grounding means through its own metallic assembly when it is used in any length not over 6 ft long and where circuit conductors within the flex are protected at not over 20 A for 3/8-in. and 1/2-in. flex and at not more than 60 A for 3/4-in., 1-in., and 1 1/4-in. flex.

2. Section 351-8 requires clamping of liquidtight flex every 4 1/2 ft and within 12 in. of each termination.
3. A 3-ft length may be used without clamping "where flexibility is necessary"—as at transformers, motors, etc.
4. A 6-ft length may be used without clamping when used for connection of a lighting fixture that has a high-temperature wiring compartment and must be wired with tap conductors of temperature rating higher than that of the branch circuit conductors—as permitted by Sec. 410-67(c).

Very important!

As noted above, both flexible metal conduit and liquidtight flex may be used without clamping—Secs. 350-4 and 351-8. In both cases, the use without clamping is permitted "at terminals where flexibility is necessary."

But, Watch Out for This! Both Sec. 350-5 and Sec. 351-9 contain an "Exception No. 2" that makes it mandatory to use an equipment grounding conductor with flex and liquidtight at any time either product is "used to connect equipment where flexibility is required." That Exception to each Section must be taken to require use of an equipment grounding conductor in each and every installation where the flex or liquidtight is not clamped within 12 in. of each termination and every 4 1/2 feet between terminations.

The safest way to assure compliance with the interrelated rules of Secs. 350-4 and 350-5 or the rules of Secs. 351-8 and 351-9 is to use an equipment grounding conductor for any unclamped run of flex or liquidtight that is used "where flexibility is necessary"—as at transformers and motor connections or for connection to other equipment that might produce vibrations or require movement. In both Sec. 350-5 and Sec. 351-9, Exception No. 2 is intended to nullify Exception No. 1 that permits use of flex or liquidtight without need for an equipment grounding conductor.

Figures 3 and 4 show actual job installations which relate to Code rules discussed here.

Figure 3 Flex connections from a generator terminal box up to the busway that carries the generator output current to the generator disconnect must always contain an equipment grounding conductor, which must be sized in the same way as described for a transformer secondary circuit. Because the output circuit of the generator has no overcurrent protective device at the terminal box, the minimum required size of the equipment grounding conductor would be selected from Table 250-95 based on the rating of overcurrent device that would properly protect the output circuit conductors at their ampacity.

When the output circuit of a generator (or the secondary circuit of a transformer) is run with conductors in parallel, as shown here, the question arises, "How do we size and run the equipment grounding conductors?" In the case shown here, if the four flex runs from the generator terminal box up to the terminal box for the busway carry a circuit made up of four 500 MCM THHN conductors per phase leg and four for the neutral, the circuit would be taken as having an ampacity of 4 × 430 A (the ampacity of a single 500 MCM copper THHN) or 1720 A. A standard 1600-A protective device would protect such a circuit, so we can use the 1600-A rating to enter Table 250-95, where we find that calls for a minimum of 4/0 copper for an equipment grounding conductor. Then because Sec. 250-95 says that a *full*-size equipment grounding conductor must be used in each of the raceways making up a parallel circuit, a 4/0 copper bare or insulated conductor could be run in each length of flex to provide the ground continuity from the generator box to the busway box—with the proper grounding terminal connections in each box. Beyond any doubt, that would satisfy all Code rules.

But another way of looking at this makeup would be to consider the generator output circuit as a service (which it is according to the Code definition of "service") and size the bonding conductors for the flex in accordance with Sec. 250-79(c) because such jumpers are on the "Supply Side of Service"—that is, the supply side of the generator disconnect, which serves as the standby or emergency service disconnect. Sec. 250-79(c) says that, for each raceway of a parallel makeup, the size of the bonding jumper for each raceway may be based on the size of the conductor in each raceway—the 500-MCM phase leg in each flex. Then Table 250-94 would call for a No. 1/0 copper in each flex run. Such application would depend upon the concurrence of any Code-enforcing authority. But use of a 4/0 conductor in each flex run, as described above, would completely satisfy all applicable Code rules.

When calculating required size of equipment grounding conductors or bonding jumpers, it is important to keep in mind the concept covered in Exception No. 2 of Sec. 250-95: "The equipment grounding conductor shall not be required to be larger than the circuit conductors supplying the equipment."

Figure 4 Nonflexible connections from a transformer to primary-side and secondary-side metal enclosures do not require an equipment grounding conductor when run in rigid metal conduit, IMC, or EMT, because all those raceways are recognized by Sec. 250-91(b) as suitable equipment grounding conductors themselves. An equipment grounding conductor (or bonding jumper) is required for flexible metal conduit, liquidtight flexible metal conduit, rigid nonmetallic conduit, and electrical nonmetallic tubing (ENT).

Outdoor PVC Conduits Need Expansion Couplings

Sec. 300-7(b). **When PVC conduit runs are installed outdoors, the variations in outdoor temperatures and the heat of the sun cause expansion and contraction in the plastic material of the conduit. If such conduit runs are not equipped with expansion couplings to accommodate changes in conduit length as a result of temperature increases and decreases, there will usually be buckling and distortion of the conduit lengths.**

Determining the actual number and the installation "pre-set" of expansion couplings involves specific calculation steps:

Step 1 For any PVC conduit run, it is first necessary to determine the change in length that a conduit run will undergo as a result of the actual temperature variation in the location of the conduit installation. For any conduit length, the number of inches of change in the conduit length can be determined from the following formula:

Inches of length increase = coefficient of thermal expansion (K) × the length of the conduit run × the temperature change from the lowest to the highest outdoor temperature that prevails at the place of the conduit installation.

Example For a 200 ft conduit run in a place where the lowest winter temperature is 0°F and the highest will be 110°F, the formula reveals the following:

$$\text{Length increase} = K \times 200\ \text{ft} \times 110°\text{F}$$

Then, with $K =$ to 3.96×10^{-4} in./ft./°F

$$\text{Length increase} = 3.96/10{,}000 \times 200 \times 110°\text{F}$$
$$= 8.71\ \text{inches}$$

Step 2 The number of expansion couplings required for the length of the conduit run to accommodate the 8.71 in. of expansion can be determined by dividing the length of the expansion (8.71 in.) by the number of inches one expansion coupling can handle (e.g., 4 in.):

NUMBER = 8.71 in. (total expansion)/4 in. (coupling travel)

NUMBER = 2.18, but 3 couplings should be used.

Then the 3 couplings can be spaced at intervals of 50 ft. between the ends of the conduit run, as follows:

END → ⟷ ⟷ ← END

50ft 50ft 50ft 50ft

Step 3 Each expansion coupling must be installed to allow both expansion and contraction of the length of the conduit run. The "pre-set" or amount of "piston opening" for any coupling at the time of installation must be determined based on the following formula to allow both expansion and contraction of the conduit lengths.

$$PO = (T_m - T_{ins}/dT) \times E$$

where PO = piston opening in inches
T_m = maximum anticipated temperature of conduit (air) in °F
T_{ins} = temperature of conduit (air) at time of installation
dT = total change in conduit (air) temperature
E = expansion allowance of each coupling in inches

Expansion coupling (arrow) is typical of such fittings that were installed at 50 ft. intervals in both the 2½ in. and 3 in. PVC conduits that run for 370 ft. on the roof of a large industrial building where summer sun and high ambient temperature would surely distort and buckle the conduit if expansion fittings were not used. The contractor making this installation had seen another job where PVC conduit had warped and buckled when a bank of six runs of 3 in. conduit had been installed across the roof of a building. Although calculations here indicated need for an expansion coupling about every 90 ft., such couplings were used every 50 ft. to allow for more expansion because of change of direction along the runs. A PVC, Type "C" conduit body (the box-like fitting at right) was installed at about 80 ft. intervals along each run to facilitate pulling of the conductors. (Four, 3/0 THHN plus No. 2 green ground in the 3 in. pipe; three, 3/0 THHN plus No. 2 green ground in the 2½ in. pipe.) After the pull-line was installed in the conduit, but before the pull was made, each "C" conduit body was filled with wire pulling lubricant to provide automatic dispensing of the lubricant to the surface of the conductors as they moved through each fitting. Conductors were pulled a certain distance until the lubricant in the fittings was used up, then the fittings were repacked with lubricant for pulling another distance, etc. This technique added great speed and ease to the pulling of these long runs.

For the **Example** shown in Step 1, with a maximum anticipated temperature of 110°F and the installation made when air temperature is 80°F, the above formula works out as follows:

$$PO = (110°F - 80°F/110°F) \times 4$$

$$= 1.09 \text{ in.}$$

Therefore, in this case, the expansion coupling must be installed with the piston extended 1.09 in. from the fully closed position.

Conductor Ampacity Depends on Conditions of Use

Sec. 310-15. **Selection of conductors for all circuits—services, feeders, branch circuits—involves carefully choosing among the many sizes of conductors and types of insulations to pick out the right conductor for the load current to be supplied. BUT, the basic element of the task depends upon a sure and confident ability to determine how much current any given conductor is permitted to carry by the 1990** National Electrical Code. **Here's the whole story.**

One of the most fundamental tasks facing everyone involved with electrical work is establishing the ampacity—the maximum Code-recognized current carrying capacity—of every circuit conductor. This article covers the basic (but somewhat complicated) procedure for determining the ampacity of *any* conductor used for a branch circuit, a feeder, a subfeeder, or a service entrance. If you do *not* have a complete and detailed understanding of this subject, your work is exposed to code violations, safety hazards, and doubtful economic validity. And because selection of fuses and circuit breakers *must* be based on the conductor ampacity (Sec. 240-3), less than a sure, confident understanding of the procedure described here creates a virtual impasse in any circuit design.

The NEC approach to establishing the ampacity of any conductor is aimed at designating the current that will cause the conductor insulation to reach and stabilize at its thermal limit, the temperature beyond which it will be damaged. The Fine Print Note (FPN) following

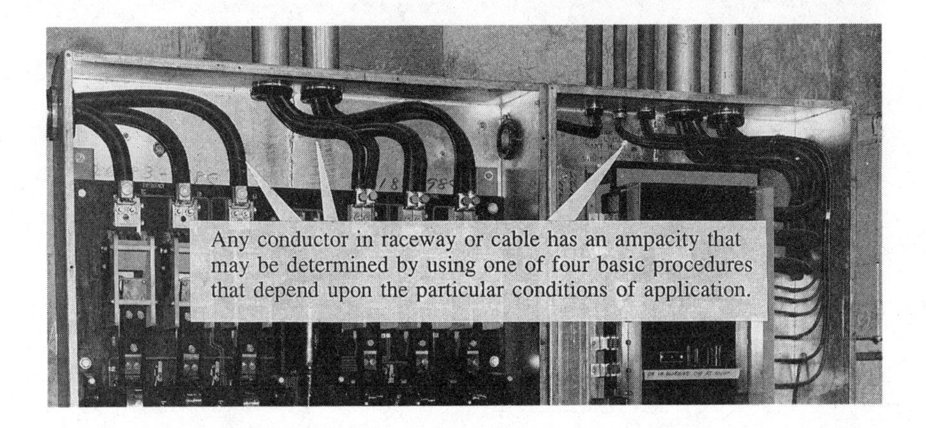
Any conductor in raceway or cable has an ampacity that may be determined by using one of four basic procedures that depend upon the particular conditions of application.

The conductors in these conduits that are run across the roof of a building will be baking in the sun on hot summer days and, therefore, require ampacity derating for elevated ambient temperature.

AMPACITY: The current in amperes a conductor can carry continuously under the conditions of use without exceeding its temperature rating.

Sec. 240-1 verifies this assessment. The wording of the first sentence here, which has been in the Code and virtually unchanged for over 35 years, reads:

> Overcurrent protection for conductors is provided to open the circuit if the current reaches a value that will cause dangerous temperature in conductors or conductor insulation.

As described in the definition of "ampacity" given in Article 100, ampacity is the amount of current, in amperes, that the conductor can carry continuously under specified conditions of use without developing a temperature in excess of the value that represents the maximum temperature that the conductor insulation can withstand. NEC Table 310-16, for instance, specifies ampacities for conductors where not more than three (current-carrying) conductors are contained in a single raceway or cable or directly buried in the earth, provided that the ambient temperature is not in excess of 30°C (86°F).

In the 1984 NEC, Tables 310-16 through 310-19 and their accompanying notes produced a clear, logical, coherent, and comprehensive approach to sizing conductors for the load current. However, in the 1987 edition of the Code, there was considerable change in the approach for determining conductor ampacity. This change was intended to recognize the use of the Neher-McGrath method of calculating conductor ampacity. The net result was a procedure that ranged from very difficult to comprehend and apply, at best, to totally incomprehensible and unenforceable, at worst. While the '90 Code still permits the use (if that's possible) of this method [as now covered in Sec. 310-15(b)], this discussion and analysis is based on the procedure presented by the rule of Sec. 310-15(a) in the 1990 NEC, which states:

> Ampacities for conductors rated 0 through 2000 volts shall be as specified in Tables 310-16 through 310-19 and their accompanying notes.

The basics

When all NEC rules on conductor ampacity are considered and correlated, it can be clearly established that any given conductor in a raceway, in a cable, or directly buried in the ground must have its ampacity determined in accordance with one of four possible conditions of use. Summarized, they are as follows:

1. When there are *not more than three* current-carrying conductors in a raceway or cable or directly buried, and the ambient temperature is *not over 30°C (86°F),* the conductor has an *ampacity equal to* the current value shown in Table 310-16 for the particular size and insulation of the conductor.
2. When there are *not more than three* current-carrying conductors in a raceway or cable or directly buried but the ambient temperature *exceeds 30°C (86°F),* the conductor has an *ampacity that is calculated* by multiplying the current value shown in Table 310-16 (for the particular size and insulation of the conductor) times the "ampacity correction factor" selected from the bottom part of Table 310-16, based on the actual ambient temperature and the particular size and insulation of the conductor.
3. When there are *more than three* current-carrying conductors in a

raceway or cable or directly buried and the ambient is *not over 30°C (86°F)* the conductor has an ampacity *that is calculated* by multiplying the current value shown in Table 310-16 (for the particular size and insulation of the conductor) times the percentage shown in the table to Note 8 to Table 310-16, based on the actual number of current-carrying conductors in the raceway, cable, or directly buried.

4. When there are *more than three* current-carrying conductors in a raceway or cable or directly buried and the ambient *exceeds 30°C (86°F)* the conductor has an *ampacity that is calculated* by multiplying the current value shown in Table 310-16 (for the particular size and insulation of the conductor) times the "ampacity correction factor" from the bottom part of Table 310-16 (based on the actual ambient temperature and the particular size and insulation of the conductor) *and then* times the percentage shown in the table to Note 8 to Table 310-16 (based on the actual number of conductors).

After the ampacity of a given conductor is established by one of the four above procedures, a suitable overcurrent protective device must be selected on the basis of that established ampacity. In accordance with all the rules of Sec. 240-3, the maximum current ratings of the protective device (fuse or CB pole) must be determined in accordance with the ampacity of the circuit conductor to be protected.

The basic rule of Sec. 240-3 requires that every conductor be protected in accordance with its ampacity—as determined above. That is, the fuse or CB pole protecting a conductor must ideally be rated at not more than the conductor ampacity. But, because ampacity values as calculated above often do not exactly match the rating of standard available fuses or CBs (as given in Sec. 240-6), Exception No. 4 to the basic rule of Sec. 240-3 says that, in such cases, the conductor may be considered to be protected by a fuse or CB that has a current rating that is the next higher standard rating of protection above the ampacity value of the conductor—**but only up to 800A.** When a conductor has an ampacity greater than 800A and there is no standard rating of fuse or circuit breaker that exactly corresponds to the ampacity value, the conductor *may not* be protected by a fuse or CB with a rating above its ampacity. The maximum rating of fuse or CB that may be used to protect such a conductor is either the next lower *standard* rating *or* (as covered in the FPN in Sec. 240-6) any listed *nonstandard* rated overcurrent protective device whose rating does not exceed the conductor's ampacity. (Remember, fuses or CBs for protecting motor-circuit conductors are permitted in Sec. 430-52 and 430-62 to be substantially higher than the conductor ampacity.)

In the 1987 edition, the NEC required that a circuit breaker with an adjustable or changeable long-time trip rating must have load-circuit conductors of an ampacity at least equal to the highest rating at which the adjustable trip might be set. But, a new Exception to Sec. 240-6 now recognizes the use of such a CB with a conductor of an ampacity less than the highest possible trip setting *if* the CB is equipped with a lockable or sealable cover; or is in a locked enclosure; or is in a locked room. In such an application, however, the actual trip setting must protect the conductor in accordance with its ampacity, as required by the basic rule of Sec. 240-3. This is more consistent with the Code-recognized treatment of fusible switches in that fusible switches still, and always could be, cabled for the actual rating of the fuse, and not the maximum value of fuse that might be installed.

From the definition of "ampacity," it is clear that heating of the conductor is the factor that determines ampacity ("without exceeding its temperature rating"). It is also clear that "ampacity" is the amount of current that the conductor can carry under *a specific set* of "conditions of use." And finally, it is also clear that when the "ampacity" of a conductor is determined in accordance with all applicable Code rules, that it is a "continuous" rating. The NEC makes no differentiation between ampacity for continuous load and for noncontinuous load. Any determined value of ampacity represents the amount of current the conductor is rated to carry continuously or noncontinuously.

After determining the maximum permitted rating of protective device for the particular ampacity of the conductor, it is then necessary to consider whether the load is "continuous" or "noncontinuous" (i.e., continuing steadily for three hours or more).

The matter of continuous vs. noncontinuous load has nothing to do with conductor ampacity because the rated ampacity of a conductor is a continuous rating and if a conductor can carry a value of current "continuously," it can certainly carry the same value of current "noncontinuously." But, the question of continuous and noncontinuous load is specifically related to application of CBs and fusible switches—some of which are suitable (UL-listed) for carrying continuous load up to 100% of their rating, while others are suitable for carrying *noncontinuous* current up to 100% of their rating BUT are *not* suitable for carrying *continuous* current (steady current flow for more than 3 hrs) that is in excess of 80% of the rating of the circuit breaker or fuse.

In selecting a CB or fuse to protect a conductor of a particular ampacity, the maximum permitted rating of the protective device is the same whether or not the load is continuous or noncontinuous. *But* the load current on the circuit must satisfy the following:

1. If a molded case CB is not specifically marked (on its label) that it is suitable for continuous loading up to 100% of its marked current

rating, then the continuous load must be limited. As given in Secs. 210-22(c), 220-3(a), and 220-10(b), the rating of the CB must be at least equal to the value of noncontinuous current plus 125% of the continuous current. At present, there are no available molded-case CBs with a 100% continuous rating in frame sizes under 600A. All of the usual branch-circuit CBs (15 to 50A) must have a rating equal to the noncontinuous load plus 125% of the continuous load.

2. Any molded-case CB rated and marked for continuous loading up to 100% of its rating may be used with circuit loading up to that level.
3. Any fusible switch that is not marked for continuous loading to 100% of its rating must have the load on its protected circuit conductors limited. The rule of Secs. 210-22(c), 220-3(a), and 220-10(b) also apply to fusible switches. *The rating of the fuse used in the switch* (not the ampere rating of the switch) must be at least equal to the total noncontinuous load plus 125% of continuous load. When a fusible switch is marked and UL-listed for 100% continuous loading (such as bolted-pressure and butt-contact pressure switches), the circuit controlled and protected by the fused switch may be loaded up to 100% of the rating of the fuse.

Typical example of applying these rules

Any given size of conductor and type of insulation used in raceway or cable has only one of four possible procedures, depending on conditions of use, to determine its ampacity. The following exercise will examine these different procedures as they apply to a conductor of specific size and insulation type and provide a basis for developing a thorough understanding of this subject. The procedures and concepts put forth in this exercise can then be applied to a conductor of any size or insulation type to determine its ampacity. Give this Code change your fullest attention and highest priority.

Question What is the ampacity of a No. 6 AWG, THW, copper conductor?

Answer It can have any one of hundreds, even thousands, of different ampacities. The ampacity will be determined by the conditions of use.

Condition 1

When the conditions of use are as described by the heading of one of the ampacity Tables, the ampacity of the conductor can be taken directly from that ampacity Table. (See accompanying illustration of Table 310-16.)

240-3. Protection of Conductors — Other than Flexible Cords and Fixture Wires. Conductors, other than flexible cords and fixture wires, shall be protected against overcurrent in accordance with their ampacities as specified in Section 310-15.

Exception No. 4: Next Higher Overcurrent Protective Device Rating. Where the ampacity of the conductor does not correspond with the standard ampere rating of a fuse or a circuit breaker without overload trip adjustment above its rating (but which may have other trip or rating adjustments), the next higher standard device rating shall be permitted only if this rating does not exceed 800 amperes and the conductor is not part of a multioutlet branch circuit supplying receptacles for cord- and plug-connected portable loads.

This is the basic table to use for over 99% of all conductor applications.

Table 310-16. Ampacities of Insulated Conductors Rated 0-2000 Volts, 60° to 90°C (140° to 194°F) Not More Than Three Conductors in Raceway or Cable or Earth (Directly Buried), Based on Ambient Temperature of 30°C (86°F)

Size	Temperature Rating of Conductor. See Table 310-13.								Size
	60°C (140°F)	75°C (167°F)	85°C (185°F)	90°C (194°F)	60°C (140°F)	75°C (167°F)	85°C (185°F)	90°C (194°F)	
AWG kcmil	TYPES †TW, †UF	TYPES †FEPW, †RH, †RHW, †THHW, †THW, †THWN, †XHHW †USE, †ZW	TYPE V	TYPES TA, TBS, SA SIS, †FEP, †FEPB, †RHH, †THHN, †THHW, †XHHW	TYPES †TW, †UF	TYPES †RH, †RHW, †THHW, †THW, †THWN, †XHHW †USE	TYPE V	TYPES TA, TBS, SA, SIS, †RHH, †THHW, †THHN, †XHHW	AWG kcmil
	COPPER				ALUMINUM OR COPPER-CLAD ALUMINUM				
18	....	....	....	14	....	....	....	....	....
16	....	....	18	18	....	....	....	....	....
14	20†	20†	25	25†	....	....	....	....	....
12	25†	25†	30	30†	20†	20†	25	25†	12
10	30	35†	40	40†	25	30†	30	35†	10
8	40	50	55	55	30	40	40	45	8
6	55	65	70	75	40	50	55	60	6
4	70	85	95	95	55	65	75	75	4
3	85	100	110	110		75	85	85	
	95	115	125				100	100	
		130					110		

NEC Table 310-16 gives ampacities under two conditions: that the raceway or cable containing the conductors is operating in an ambient temperature not over 30°C (86°F) and that there are not more than three current-carrying conductors in the raceway or cable. Under those conditions, the ampacities shown correspond to the thermal limit of each particular insulation and the value shown *is* the ampacity of each conductor.

Use these reduction factors to calculate ampacity when ambient temperature is above 86°F.

1500	[illegible]	625	680	[illegible]	[illegible]	520	56[illegible]	[illegible]	[illegible]
1750	545	650	705	735	[illegible]	545	595	[illegible]	1750
2000	560	665	725	750	470	560	610	630	2000
AMPACITY CORRECTION FACTORS									
Ambient Temp. °C	**For ambient temperatures other than 30°C (86°F), multiply the ampacities shown above by the appropriate factor shown below.**								**Ambient Temp. °F**
21-25	1.08	1.05	1.04	1.04	1.08	1.05	1.04	1.04	70-77
26-30	1.00	1.00	1.00	1.00	1.00	1.00	1.00	1.00	79-86
31-35	.91	.94	.95	.96	.91	.94	.95	.96	88-95
36-40	.82	.88	.90	.91	.82	.88	.90	.91	97-104
41-45	.71	.82	.85	.87	.71	.82	.85	.87	106-113
46-50	.58	.75	.80	.82	.58	.75	.80	.82	115-122
51-55	.41	.67	.74	.76	.41	.67	.74	.76	124-131
56-60	....	.58	.67	.71	....	.58	.67	.71	133-140
61-70	....	.33	.52	.58	....	.33	.52	.58	142-158
71-80	....	....	.30	.41	....	....	.30	.41	160-176

† Unless otherwise specifically permitted elsewhere in this Code, the overcurrent protection for conductor types marked with an obelisk (†) shall not exceed 15 amperes for 14 AWG, 20 amperes for 12 AWG, and 30 amperes for 10 AWG copper; or 15 amperes for 12 AWG and 25 amperes for 10 AWG aluminum and copper-clad aluminum after any correction factors for ambient temperature and number of conductors have been applied.

The Ampacity Correction Factors, which must be applied to the value given in the ampacity table when the conductors are to be installed where the ambient temperature is in excess of 86°F (30°C), appear at the bottom of each of the ampacity tables. Derating of the table values is required to compensate for the diminished heat-dissipating capability of the conductors and prevent them from reaching an operating temperature in excess of that for which the conductor insulation material is rated.

This data must be applied to determine conductor ampacity when more than three current-carrying conductors are installed in a single raceway or cable.

Number of Conductors	Column A Percent of Values in Tables as Adjusted for Ambient Temperature if Necessary	Number of Conductors	Column B** Percent of Values in Tables as Adjusted for Ambient Temperature if Necessary
4 through 6	80	4 through 6	80
7 through 9	70	7 through 9	70
10 through 24*	70	10 through 20	50
25 through 42*	60	21 through 30	45
43 and above*	50	31 through 40	40
		41 through 60	35

* These factors include the effects of a load diversity of 50 percent.
**No diversity.

Whenever there are more than three current-carrying conductors in a raceway, a cable, or directly buried, the basic rule of Note 8 requires derating in accordance with the percentages given in either Column A or B of this table, depending on conductor "diversity." The whole question of what is meant by "diversity" is still unresolved because there is no definition given for this term anywhere in the NEC.

As a result of the changes in the 1990 NEC, there are now only four tables covering ampacities for copper, aluminum, and copper-clad aluminum conductors rated 0 to 2000V. Tables 310-16 and 310-17 are for conductors with insulation rated from 60 to 90°C in an ambient temperature no greater than 86°F (30°C) and Tables 310-18 and 310-19 cover conductor insulations rated from 150 to 250°C in an ambient temperature of 104°F (40°C). Both Tables 310-16 and 310-18 are for use where there are not more than three current-carrying conductors in a raceway, cable, or directly buried, and Tables 310-17 and 310-19 are for "single insulated conductors...in free air" or, in other words, spaced open conductors. Inasmuch as the vast majority of electrical circuits incorporate 60 to 90°C-rated insulated conductors in raceway, cable, or directly buried, it quickly becomes clear that Table 310-16 is going to be used for the vast majority of installations, including our example.

Although Tables 310-16 through 310-19 refer to "*three conductors* in raceway...," it should read "three *current-carrying* conductors." There are applications where more than three conductors are installed in a raceway, or cable, or directly buried, BUT where all of the conductors are *not* counted because the Code does not consider them to be current-carrying conductors. For example, a 3-phase, 4-wire, branch circuit with an equipment grounding conductor, consists of five conductors, but in many instances, not more than three of the conductors are considered to be current-carrying conductors. As covered by Note 10, the neutral in such a circuit "shall not be counted" as one of the "three conductors" referred to in the Tables' headings, if less than half the load is electric discharge lighting, data processing equipment, or other "nonlinear" loads. And Note 11 to the Ampacity Tables, states that a grounding or bonding conductor also "shall not be counted." Therefore, even though there are actually five conductors, only three "count" as current-carrying conductors, and the ampacity can be taken directly from the appropriate Table for the particular size and insulation of the conductor.

As shown in Fig. 1, the ampacity of each No. 6 AWG, THW, copper conductor [where there are not more than three current-carrying conductors in the raceway and the ambient temperature is not in excess of 86°F (30°C)—i.e., the conditions set forth in the heading of Table 310-16] is taken directly from the table. Under these conditions of use, the ampacity shown in the table corresponds to the thermal limit of this particular insulation, and the ampacity of each No. 6 AWG, THW, copper conductor is 65A.

Rating of overcurrent protection. Because there is not a standard rated fuse or circuit breaker of 65A rating, each No. 6 AWG, THW, copper

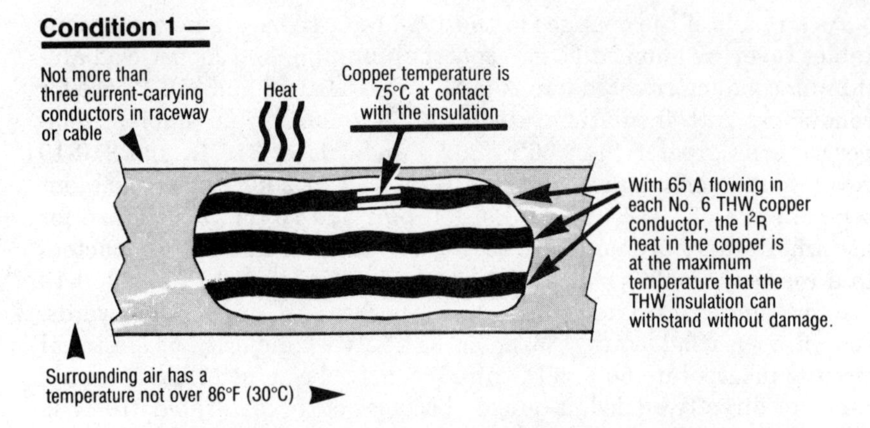

Under these conditions, No. 6 THW copper conductor has the ampacity value of *65 A*, as shown in Table 310-16.

Figure 1

conductor may be protected by the next standard rating of protective device above 65A [Exception No. 4 to Sec. 240-3]. Sec. 240-6 shows that to be a 70A pole or fuse.

The maximum Code-recognized loading permitted on each No. 6 THW conductor would be 65A—if the load current does not persist for a period of 3 hours or more. That is, if the loading does *not* constitute "continuous load" (where the current flows for 3 hours or more, as defined in NEC Article 100 on "Definitions"), the circuit may be properly loaded up to the conductor ampacity of 65A—*BUT* not over 65A, even though the 70A protection would permit current higher than 65A.

OR

56A—if the current loading is "continuous load" (3 hours or more) in which case the loading would be limited to not more than 80% of the 70A fuse or CB—as required by UL rules that limit continuous loading to not more than 80% of the rating of a fuse or circuit breaker [unless the fuse or CB is UL listed for continuous loading to 100% of its rating, and up to 400A, there are *none* so listed]. And that same limitation for continuous loading is also required by the rules of Sec. 210-22(c) and 220-3(a)—both of which require that a branch circuit protective device be rated *not less than* 125% of value of a continuous load. Using the 70A protection, the *maximum* permitted continuous load would be 70 divided by 1.25 or 56A. Sec. 220-10(b) would set the same maximum of 56A continuous load if the circuit in Condition 1 was a feeder instead of a branch circuit. Note that by multiplying continuous load current by 125% to obtain the minimum rating of protective device is the reciprocal of multiplying the rating of the protective device by 80% to obtain the maximum value of continuous load.

Condition 2

NEC Table 310-16 gives ampacities under two conditions: that the raceway or cable containing the conductors is operating in a surrounding ambient temperature not over 86°F (30°C) and that there are not more than three current-carrying conductors in the raceway or cable. Under those conditions, the ampacities shown correspond to the thermal limit of each particular insulation and the value shown *in the Table is* the ampacity of each conductor. But, in any case where either (or both) of the two conditions is exceeded, the ampacity of the conductors must be reduced (and protection must be based upon or provided at the reduced ampacity!) to ensure that the temperature limit of the insulation is not exceeded.

If the ambient temperature is above 86°F (30°C), the current value given in Table 310-16 for the given size and insulation of the conductor must be reduced by the appropriate correction factor from the bottom part of Table 310-16, for the particular temperature that exists. (See accompanying table of "Ampacity Correction Factors" for ambient derating.) However, while that requirement is easy to understand, the Code provides no guidance on how to determine an appropriate adjusted "ambient temperature." Should it be the maximum temperature reached? An average maximum? The Code does not say.

In the absence of any direct guidance from the NEC, one approach that should produce satisfactory results in actual application would be to determine a 3-hour, "average maximum" over the course of a week. Ideally (but realistically improbable), hourly temperature readings would be taken over an entire day to determine that 3-hour period when the ambient temperature is greatest. Then for the next week, take hourly readings over this 3-hour period. Average these readings and use that temperature value for selecting the appropriate derating factor from the "Ampacity Correction Factors" for ambient temperature that appear at the bottom of Tables 310-16 through 310-19. While such an approach is not always practicable, especially at the design stage, the only other method available to determine what the adjusted ambient temperature should be is to take an "educated guess" at an "average maximum" based on previous experience.

In Fig. 2, it can be seen that while there are still only three current-carrying conductors within the raceway, each No. 6 AWG, THW, copper conductor now has an ampacity of 53A. This is the product of multiplying the ampacity value from Table 310-16 (65A) times the factor

Condition 2 —

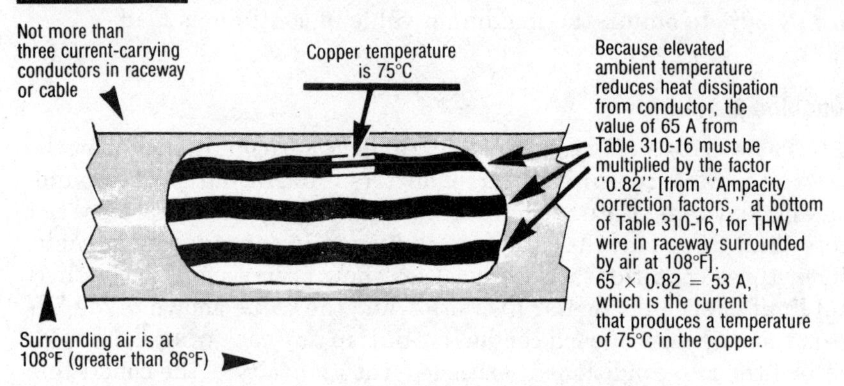

Under these conditions, No. 6 THW copper conductor has an ampacity value of *53 A,* and not 65 A as shown in Table 310-16.

Figure 2

(0.82) from ampacity correction table for an ambient temperature of 108°F, which expressed mathematically would be 65A × 0.82 = 53A. Because these conductors are in an elevated ambient, they are less able to dissipate heat. Under these conditions, 53A now represents the conductor ampacity—that is, the amount of current that will produce an I^2R heat generation, establishing thermal equilibrium at a temperature of 75°C in the copper, which is the maximum temperature that THW insulation can withstand without damage.

Rating of overcurrent protection. Here, each No. 6 AWG, THW, copper conductor, with an ampacity of 53A, may be protected by a fuse or breaker rated at 60A, which is the next standard rating of protective device above the conductor's 53A ampacity—again, as permitted by Exception No. 4 to Sec. 240-3, which permits protection by the next standard rating of protective device above the conductor's ampacity when there is not a standard rated protective device of the conductor's ampacity.

The maximum Code-recognized loading permitted on each No. 6 conductor would be 53A—that is, up to the conductor's ampacity—if the loading is "noncontinuous load."

OR

48A—which is 80% of the 60A rating of the protective device—if the total load on the circuit will persist for any period of 3 hours or more, that is, for "continuous load." The same UL rule and Code sections described above would apply to limit continuous load to not more than 80% of the rating of the fuse or CB protecting the circuit.

Condition 3

When there are more than three current-carrying conductors in a raceway or cable, their current-carrying capacities must be decreased to compensate for proximity heating effects and reduced heat dissipation due to reduced ventilation of individual conductors, which are bunched or which form an enclosed group of closely placed conductors. In such cases, the ampacity must be reduced from the Table 310-16 value, as required in the table of Note 8, which appears after Tables 310-16 through 310-19.

Part (a) of Note 8 to Table 310-16 says: "Where the number of conductors in a raceway or cable exceeds three, the ampacities given shall be reduced as shown in the following table."

This table was modified in the 1987 edition and further revised in the 1990 NEC. As a result there are now two columns of derating factors from which to choose based on the "diversity" of the conductors. As given in the footnotes to the table, Column A is based on a "diversity of 50 percent" and Column B is based on "no diversity." Because the Code has no definition for "diversity," it is virtually impossible to defend the use of factors given in Column A with any degree of certainty. Therefore, the safest, easiest, and most defensible approach would be to always use the factors given in Column B whenever derating for more than three current-carrying conductors in a raceway, or cable, or directly buried.

Fig. 3 shows that each No. 6 AWG, THW, copper conductor now has an ampacity of 52A. This is the product of multiplying the ampacity value from Table 310-16 (65A) times the factor (0.80) for 4 through 6 conductors given in Column B of the table to Note 8 (65A × 0.80 = 52A). Because more than three current-carrying conductors are used in a single raceway or cable, their heat-dissipating capability is reduced. Under these conditions, 52A now represents the conductor ampacity—that is, the amount of current that will produce an I^2R heat generation, establishing thermal equilibrium at a temperature of 75°C in the copper, which is the maximum temperature that THW insulation can withstand without damage.

Rating of overcurrent protection. As in Condition 2, a 60A fuse or CB pole may be used to protect each No. 6 AWG, THW, copper conductor with its ampacity of 52A, for the same reasons explained under Condition 2. And the maximum Code-recognized loading would be 52A—up to conductor ampacity—for a "noncontinuous load."

Condition 3 —

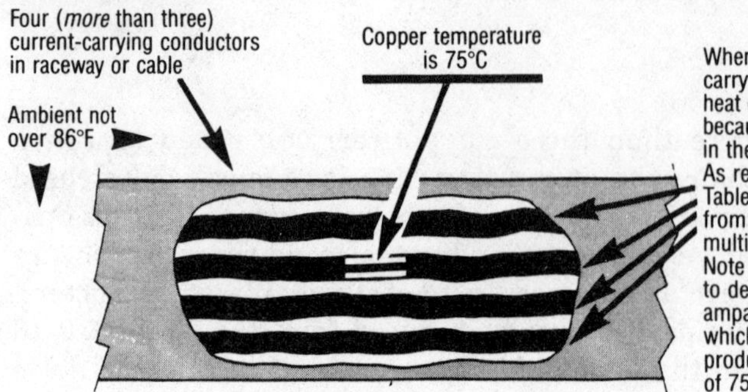

Under these conditions, No. 6 THW copper conductor has an ampacity value of 52 *A*—not 65 A.

Figure 3

OR

48A—for a "continuous load," which is again 80% of the 60A protective device.

Condition 4

Where more than three current-carrying conductors are used in a single raceway or cable, AND the conduit or cable containing the conductors is in an ambient air temperature higher than 86°F (30°C), the ampacity of the contained conductors must be reduced in accordance with the factors of Note 8 AND in accordance with the "Ampacity Correction Factor" table for higher ambient temperatures given at the bottom of Table 310-16. This is made clear by the headings of Column A and B in the Table to Note 8. These headings require that any ampacity derating for elevated ambient must be made **in addition** to the one for number of conductors.

As shown in Fig. 4, each No. 6 AWG, THW, copper conductor now has an ampacity of 39A. The heading for Column B in Note 8 requires that the table value first be adjusted for ambient temperature. As was done under Condition 2, the base ampacity of the No. 6 AWG, THW, copper conductor is multiplied by the "ampacity correction factor" for an elevated ambient—that is, over 86°F—65A × 0.75 = 49A. Then this product is multiplied by the factor given in Column B for 4 through 6 conductors—49A × 0.80 = 39A. Therefore, under these conditions of use, each No. 6 AWG, THW, copper conductor has an ampacity of 39A.

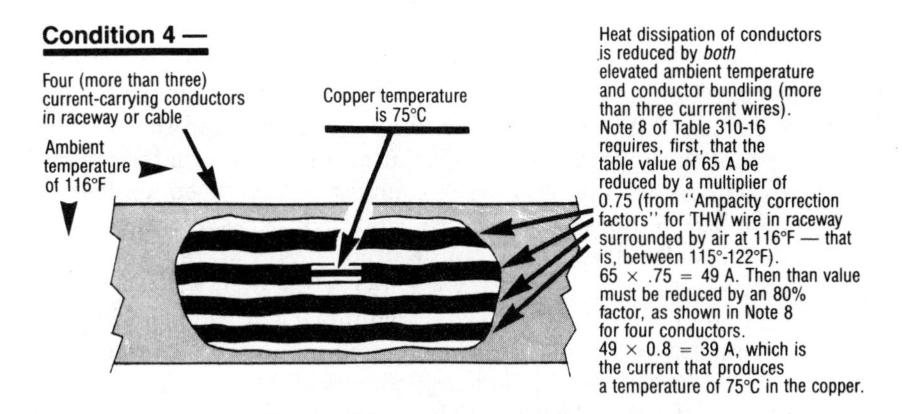

Figure 4

Rating of overcurrent protection. Now, the same No. 6 AWG, THW, copper conductors have an ampacity of only 39A—which calls for a maximum 40A rated protective device, with a loading of 39A maximum or "noncontinuous load"

OR

32A (80% of the 40A device rating) if the load is "continuous."

There are no other NE Code rules or considerations that apply to *determining* the amount of current that any conductor is permitted to carry.

Notes

It should be clearly understood that any reduced ampacity, required because of higher ambient and/or conductor bundling, has the same meaning as the value shown in Table 310-16: each represents a current value above which excessive heating would occur under the particular conditions. And if two separate conditions individually diminish the ability of the conductor to dissipate heat, then more reduction of current is required than if only one such condition existed.

In conductor sizes Nos. 14, 12, and 10 AWG, Table 310-16 clearly indicates that 90°C-rated conductors do, in fact, have higher ampacities than those given for the corresponding sizes of 60°C and 75°C conductors. As shown, for copper, No. 12 TW and No. 12 THW are both assigned an ampacity of 25A under the basic conditions of the table. But, a No. 12 THHN, RHH, or XHHW (dry location) has an ampacity of 30A. However, the obelisk (dagger) note to Table 310-16 requires that overcurrent protection for No. 14, No. 12, and No. 10 be taken as 15, 20, and 30A, respectively, regardless of the type and temperature rating of the insulation on the conductors. And the obelisk note says that these limitations apply "after any correction factors for ambient temperature and number of conductors have been applied."

When applied to the selection of branch-circuit wires in cases where conductor ampacity derating is required by Note 8 of Tables 310-16 through 310-19 for conduit fill (over three wires in a raceway), the obelisk note to Table 310-16 affords advantageous use of the higher current-ratings of 90°C wires for branch-circuit makeup. The reason is that, as stated in Note 8, the derating of ampacity is based on taking a *percentage of the actual current value shown in the table,* and the table current values for 90°C conductors *are higher* than those for 60°C and 75°C conductors. The derated ampacity must be properly protected by the 15, 20, and 30A protective devices.

Types of Cable Permitted in Cable Tray

Secs. 90-4, 318-3. **Sec. 318-3 of the NEC regulates the kind of circuit wiring that may be installed in cable tray installations. Because Type NM cable (Romex, Article 336), Type UF cable (underground feeder cable, Article 336), and Type TC cable (tray cable, Article 340) are permitted to be used in tray, designers frequently specify, for use in tray, cables that are not those types but do, in fact, have superior dielectric, chemical, and mechanical strengths. And inspectors, exercising the authority granted them in Sec. 90-4 (second paragraph), "waive" the specific cable requirements of Sec. 318-3(a) because the alternative cables do exceed the safety objectives and characteristics of the cable types enumerated in Sec. 318-3(a).**

One of the most important and realistic rules in the NEC is the second paragraph of Sec. 90-4, which authorizes any electrical inspector to permit disregard of the exact letter of any Code rule if in the inspector's judgment—from personal experience, competence, and expertise—the safety objectives of the specific Code rule are satisfied by an alternative material or method. That is a very sound and necessary rule in the NEC for the simple reason that the framers of any Code rule—the Code panel members—are attempting to set out a safety concept that would have broad general application to the vast majority of usual conditions. But it is impossible for any rule to be worded in such a way as to include alternative products or techniques that would afford an equal (or higher) level of safety. The use of "Exceptions" to Code rules is one way to deal with that problem in specific details. The second paragraph of Sec. 90-4 is a blanket acknowledgment that most Code rules admit of other ways to provide the same safety.

Although Sec. 318-3(a) accepts Types NM, UF, and TC cables in tray, that section does not specifically recognize Type SO flexible cable for use in tray. Yet there are very high quality SO cables available that could readily pass all the UL tests that are required for listing of a cable as Type NM, UF, or TC. (See Figure 1.) But the very construction that gives such SO cable electrical, chemical, and mechanical superiority to NM, UF, or TC is expensive and prices such SO cable far above the price of the other cables. And because of the substantial premium in price, manufacturers do not, for instance, submit the SO cable to UL for listing as Type TC. The manufacturer's thinking is that

Figure 1 Nonmetallic jacketed SO cable used in cable tray may be accepted by electrical inspectors in accordance with the second paragraph of Sec. 90-4 if the inspector is satisfied that the high-quality construction and jacketing of premium grade SO cable provides electrical, chemical, and mechanical performance that is equal to or superior to those characteristics of the cable assemblies specifically recognized for tray in Sec. 318-3(a)—such as Types NM, UF, or TC cable.

any user who needs TC cable will buy just that and would not pay the premium price for a superior SO-TC cable. As a result, the cost of getting the SO cable listed also as TC is an unnecessary expense.

In a recent industrial application of cable tray on an outdoor earth-handling structure, the electrical designers chose a high-quality SO cable for use in the trays because of the numbers and sizes of conductors needed for the many branch circuits for motors, lighting, and control. For the very long lengths of different multiconductor cables—requiring 4-, 6-, 8-, 10-, 20-, and 36-conductor cable assemblies of No. 14, No. 12, and No. 10 copper conductors—Type SO cable was the only readily available solution to the problem. And because many parts of the runs were subjected to movement and swaying, and all of it was constantly subjected to vibration due to the large earth-moving and conveyor motors, there was need for the extra flexibility of conductors of 41, 65, and 104 strands. From a practical real-life perspective, that need also dictated high-flexibility SO cable.

Relying on Sec. 90-4, the electrical inspector on that job permitted use of the SO cable as at least equal to TC cable, which Sec. 318-3(a) specifically recognizes. The inspector did require that the cable be suitable for wet locations and have an age-resistant outer jacket. The inspector also required that at all terminations the cable assemblies had to be marked "W–A"—for "Weather resistance" and "Age resistance"—to alert maintenance personnel that replacement of the cables must be with cable assemblies of at least the same physical characteristics. (See Figure 2.) The "SOW–A" marking indicates that the cable has passed the UL Weatherometer Test and is resistant to weather, wetness, sunlight, ultraviolet, and aging.

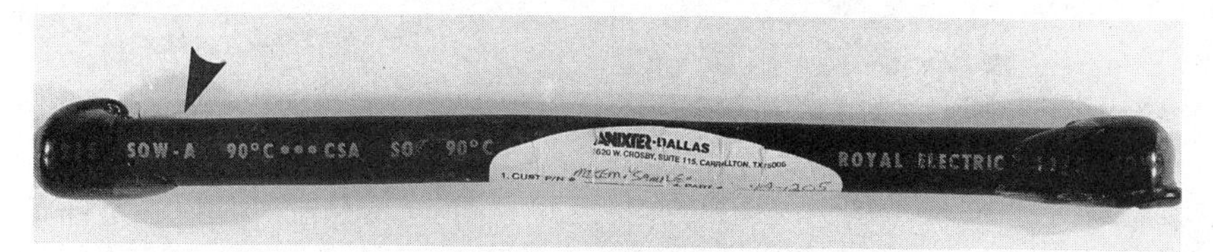

Figure 2 Sample of cable that was accepted by the inspection authorities for use in cable tray outdoors was required to have the indicated marking, "SOW–A" (arrow), by the manufacturer, indicating that the cables are suitable for exposure to weather (wetness and sunlight) and are resistant to aging.

Type UF Cable Aboveground Outdoors?

Article 339. What about the application of multiconductor Type UF cable for outdoor branch circuits and feeders when used aboveground, attached to building surfaces, or attached to poles or other structures?

As covered in NEC Sec. 339-1, Type UF cable is an "Underground feeder and branch-circuit cable" that is listed by UL and made available by manufacturers for use in accordance with NEC Article 339. Section 339-3(a) (1) clearly recognizes such cable "for use underground, including direct burial in the earth." And Sec. 339-3(a) (4) permits Type UF cable to be used "for interior wiring in wet, dry, or corrosive locations." But, Sec. 339-3(a), which covers "Uses Permitted" for UF cable does not specifically permit use aboveground outdoors; and Sec. 339-3(b), covering "Uses Not Permitted," does not prohibit outdoor use aboveground. Because the Code wording omits any direct reference to use of Type UF cable aboveground outdoors, such application frequently becomes a controversial matter with electrical inspectors.

In resolving this issue, it should first be noted that use of UF cable aboveground outdoors most certainly does **not** violate any rule of the NEC. And an electrical inspector would therefore be unable to cite a violation of a specific Code rule. An indirect approval of UF cable aboveground outdoors may be taken from part (9) of Sec. 339-3(b), which prohibits UF cable "exposed to direct rays of the sun, UNLESS IDENTIFIED AS SUNLIGHT-RESISTANT." Clearly that reference **permits** UF aboveground outdoors (which is where it would be "exposed to the direct rays of the sun") if the UF cable is marked SUNLIGHT-RESISTANT in accordance with UL testing of the ability of the cable jacket to withstand any photochemical deterioration due to the ultraviolet radiation in sunlight. The UL covers this point in its *Electrical Construction Materials Directory* (the Green Book) and in *General Information (the White Book) under the heading "Underground Feeder and Branch Circuit Cable."*

Based on the foregoing analysis, inspectors do accept the use of UF cable aboveground outdoors—as shown in Figure 1.

Figure 1 UF cable branch circuit (arrow) is clamped to the outdoor surface of this building from the nonmetallic disconnect switch enclosure at right to the A/C condenser unit at left. This cable (and all UF cable in general) is UL listed and marked as a "SUNLIGHT-RESISTANT" assembly that makes it suitable for such outdoor applications.

Electrocution Emphasizes Need for Grounding Metal Cover on In-Ground, Outdoor Junction Box

Sec. 370-18(c). **This** Code **rule requires that metal covers on "all pull boxes [and] junction boxes" must be grounded by connection to an equipment grounding conductor installed in the box, whenever such a metal cover is "within 8 feet (2.44 m) vertically or 5 feet (1.52 m) horizontally of ground...and subject to contact by persons," as required by Sec. 250-42(a).**

A recent electrocution accident focuses on the critical importance of grounding of all metal parts of outdoor enclosures that contain energized electrical parts. A law suit resulting from that electrocution emphasizes need for grounding of metal covers of outdoor junction boxes set in ground, flush with ground surface.

On a drizzling summer day, two young girls were waiting for a bus at an outdoor bus terminal area. They had taken their shoes off to frolic in the cooling rain mist. One of the girls stepped onto the metal cover of one of the flush underground junction boxes—with one foot on the cover and the other foot on the ground adjacent to the cover. Because of the presence of a voltage on the metal cover, the girl was electrocuted.

Subsequent inspection revealed that the insulation on a splice in the box had been worn away by friction from the underside of the metal cover—apparently caused over a period of time by vibrations of bus movement—and an energized conductor (277 V to ground) was in contact with the metal cover. When the girl stepped on the cover, electric shock current passed through her body and killed her.

The investigation showed that the metal cover was set on a concrete in-ground junction box to which a number of conduits connected, and although an equipment grounding conductor was run in the conduits with the circuit conductors feeding lighting poles in the area, the equipment grounding conductor was not connected to the metal cover. (See Figure 1.) If the equipment grounding conductor had connected to the cover, as soon as the insulation failure on the splice brought the 277-V conductor into contact with the cover, there would have been a direct phase-to-ground (neutral) short and the circuit breaker for that

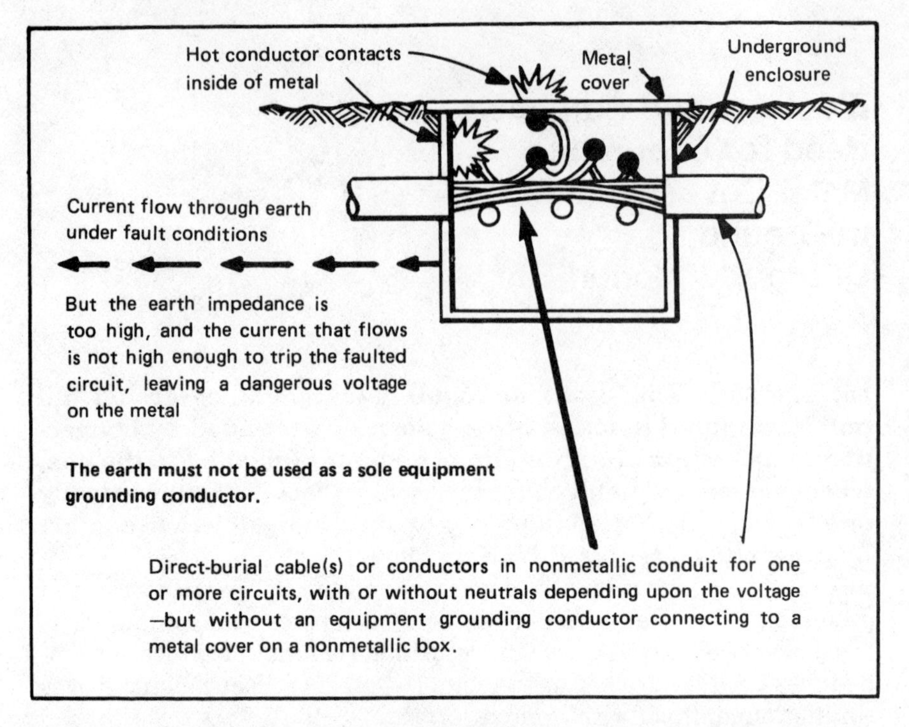

Figure 1 A metal cover on an in-ground flush junction or pull box must be properly connected to an equipment grounding conductor run with the circuit conductors coming into or through the box. As shown in this diagram, failure to make such a connection presents the hazardous possibility that any kind of an insulation failure (abrasion, cutting, nicking, or melting) on one of the energized conductors in the box will bring the conductor metal in contact with the metal of the box or of the cover, thereby placing a dangerous shock voltage on the exposed metal cover of the box. The metal box or the metal cover cannot be considered effectively grounded in accordance with Sec. 250-51 simply because the metal box or metal cover is in contact with earth (the soil) because, as Sec. 250-51 warns, "The earth shall not be used as the sole equipment grounding conductor." Therefore, connection to a ground rod would not satisfy the requirement and would be a waste of time and money. There must be a connection to a specific equipment grounding conductor, sized from Table 250-95 based on the rating of the largest overcurrent device protecting any of the circuit conductors entering the box.

circuit in the supply panelboard would have tripped open instantaneously to open the circuit and remove the hazardous condition. That type of automatic clearing of an accidental ground fault is precisely intended by Article 250 of the NEC and would have prevented the young girl's death.

Sec. 370-18(c) certainly does apply in this case—requiring grounding of the metal cover of the box, precisely to avoid the very accident that occurred in this case. But this unfortunate accident raises some important considerations that must be addressed and relate directly to the safety of the general public:

1. All over the nation, there are many thousands of outdoor in-ground electrical enclosures (containing electrical conductors) that are protected by metal covers. Such enclosures may range in size all the way from large manholes, into which one or more persons may enter and work, to very small junction boxes for branch-circuit wiring for outdoor lighting or receptacles. In-ground boxes are commonly nonmetallic boxes (concrete or plastic) with metal covers. Or they may be metal boxes with metal covers. Of course, the big concern is for exposed metal surfaces which might become energized and with which people can readily come into contact. Outdoor applications are particularly susceptible because they are exposed to the general public, especially children.
2. In manholes, the conductors are usually at the bottom of the enclosed area and are not likely to be long enough to reach up to contact the metal cover of the manhole. And even if the conductors are long enough, they would generally be secured in place and their weight would prevent vertical positioning to come into contact with the metal cover. For those reasons, metal covers on electrical manholes are not provided with connection to an equipment grounding conductor. However, the literal wording of Sec. 370-18(c) in conjunction with Sec. 250-42 could very certainly be taken to require grounding of metal manhole covers when such underground enclosures are used for pulling or splicing electrical circuit conductors.
3. But for manufactured in-ground metal, plastic, or concrete boxes that are used as required by Sec. 300-15(a), for conductor splice points or pull points for raceways, there is absolutely no doubt that a metal cover on any such box must be connected to an equipment grounding conductor run through the box.
4. Many questions arise about the correct way to provide grounding of metal covers for in-ground junction and pull boxes:
   a. Certainly if an in-ground *metal* box is properly grounded by connection to an equipment grounding conductor run with the circuit wires to the box, a metal cover on the box can be considered grounded it if is tightly screwed or bolted or otherwise connected to the grounded metal box. But use of a grounding pigtail from the equipment grounding conductor to the metal cover would be an important safety extra in the event that the screws holding the cover to the box became loose or corroded or are inadvertently not replaced after some required access to the wiring in the box.
   b. Nonmetallic boxes must have their metal covers connected somehow to the circuit equipment grounding conductor. If the nonmetallic box has a metal support rim to which the metal cover is to be attached by screws, the metal cover can be con-

sidered grounded by its attachment to the rim *if* the metal rim is connected to the equipment grounding conductor. Again though, the possibility of losing the cover grounding through its attachment to the rim would dictate a direct pigtail from the cover itself to the equipment grounding conductor.

5. In all cases where a metal cover or an in-ground box is to be grounded by a direct pigtail from the cover to the equipment grounding conductor, some liberties must be taken with the general NEC and UL prohibitions against altering or modifying equipment in the field. Some type of connection must be devised to attach the grounding pigtail to the cover. For instance, a spade-lug or eye-lug on the end of the grounding pigtail could be attached by a bolt, lockwasher, and nut to a hole drilled through the cover. The length of the pigtail can be cut to allow enough slack to enable the cover to be removed and placed back from the box. Then, in the box, the other end of the grounding pigtail is spliced to the equipment grounding conductor. The size of the grounding pigtail must not be smaller than the equipment grounding conductor indicated by Table 250-95, based on the rating of the largest fuse or circuit breaker protecting any of the circuit conductors run through the box. Use of a solidly connected grounding pigtail long enough to give enough slack in removing the cover offers the highest level of maintained safety for grounding of metal covers of in-ground splice, junction, and/or pull boxes.

Although some aspects of this important grounding application are not detailed in the NEC, the need to ground all exposed metal parts of equipment or enclosures is positively a mandatory Code rule, with the letter and the intent spelled out in Sec. 370-18(c), Sec. 250-42, Sec. 250-1 and its notes, and in many other sections of Article 250. And, as noted in the last sentence of Sec. 250-51, use of a ground rod to ground the box and/or cover would not be acceptable because current flow through the rod would have to flow through the earth and, as noted, "The earth shall not be used as the sole equipment grounding conductor."

Attachment Plugs and Receptacles Used as Motor Disconnects

Sec. 430-109, Ex. 5. To avoid accidents and liability exposure, study this critical safety data on the use of plug-caps and receptacles as the Code-recognized disconnect for the tens-of-thousands of cord-connected motors in use today.

In the 1978 NEC, as well as many previous editions, Exception No. 5 to Sec. 430-109 recognized the use of an attachment plug and receptacle as the Code-required disconnect for "portable motors." During the Code cycle for the 1981 edition of the NEC, a proposal was submitted to limit the application of this permission to motors rated not over 2-hp and operating at 300-volts or less. Although this proposal was "Accepted as revised," the revised wording did not attempt to place limits on the types of motors, but rather, on the types of receptacles and attachment plugs that would be recognized. The new wording in the 1981 NEC actually extended the permission given by Sec. 430-109, Exception No. 5 to *all* cord-and-plug connected motors—not just "portable motors" as was previously stated—*provided* that the plug and receptacle had "ratings no less than the motor ratings." This change in wording on the part of Code Making Panel (CMP) 11 caused some discussion and left some doubt as to exactly what was required for compliance with the "letter" and "intent" of the new wording used in the 1981 edition of the NEC.

Some indication of what was required was given in the wording of the second sentence of Exception No. 5, which was added by the CMP. It stated that a plug and receptacle used as the disconnect for an appliance or an air conditioner installed in accordance with other specific sections of the Code need not be "horsepower rated." Some argued that this reference indicated it was CMP 11's intent to require a horsepower-rated receptacle and plug whenever the permission given in the first sentence was exercised. Others did not agree.

Because the first sentence of Sec. 430-109, Exception No. 5 in the 1981 NEC did not specifically call for "horsepower-rated" attachment plugs and receptacles, many people felt that it was not the intent of CMP 11 to require attachment plugs and receptacles to be so rated. In addition, there were virtually no plugs and receptacles commercially available that were so listed and marked, especially in the lower ratings (i.e., 60A or less).

This point of controversy was addressed during the Code-cycle for the 1987 NEC when a proposal was submitted to include the wording "horsepower rated" in the first sentence of Exception No. 5 to Sec. 430-109 before the word "attachment plug." The substantiation indicated the purpose of this change was to clarify that it is CMP 11's intent to require such rating. And this proposal was unanimously accepted.

Once this wording was adopted and it became absolutely clear that *only* horsepower-rated attachment plugs and receptacles would be permitted in such applications, the National Electrical Manufacturers Association (NEMA) and Underwriters Laboratories, Inc. (UL) responded to eliminate the obvious void between what the Code required and what was commercially available.

Establishing horsepower ratings

Sometimes, when the NEC requires equipment to be specifically listed for a particular application, in order to allow the manufacturers an opportunity to get their products so listed, a CMP will give a "grace period" by tying compliance to an "effective date." For example, in the 1987 NEC, Sec. 820-15, which dealt with the fire-resistance of coaxial cables, mandated listing of these cables as "being resistant to the spread of fire." But, this requirement was to become effective July 1, 1988. As a result, use of coaxial cables that were not listed as "resistant to the spread of fire" was permitted for almost two years after the 1978 NEC was published.

However, this was not the case with the required horsepower-rating for attachment plugs and receptacles used as the Code-recognized disconnecting means for cord-and-plug connected motors. It would appear that because CMP 11 had intended such attachment plugs and receptacles to be horsepower rated since the 1981 edition of the NEC, no "effective date" was given when the clarified wording was adopted in the 1987 NEC.

Recognizing that immediate action was necessary, NEMA and UL agreed that developing a method for evaluating existing, listed attachment plugs and receptacles would be the better way to provide products to satisfy this requirement and thereby quickly fulfill the needs of electrical designers and installers.

Members of the NEMA Wiring Device Technical Committee, working closely with UL, developed and conducted extensive tests to determine and establish horsepower ratings for a variety of straight blade and locking-type attachment plugs and receptacles. In the NEMA straight blade configurations, the attachment plugs and receptacles evaluated were rated from 15 to 60A, 277VAC or less. For the NEMA locking-type configurations, only attachment plugs rated from 15 through 30A, 480VAC or less were tested.

After reviewing the test data, UL determined that specific horsepower ratings could be assigned to the evaluated attachment plugs and receptacles. The specific values assigned are shown in Fig. 1.

Essentially, the assigned ratings are equal to the maximum AC horsepower rating that would have a full-load current of 80% or less of the device's ampere rating.

For example, a 20A, 125/250VAC attachment plug has two horsepower ratings depending on whether the load is 125VAC (line-to-neutral) or 250VAC (line-to-line). As can be seen in Fig. 1, when the load is supplied at 125VAC, then the horsepower rating of this device

Ampere Rating	AC Voltage Rating	Phases	Poles	Horsepower Rating	NEMA Designation
15	125	1	2	1/2	1-15, L1-15 5-15, L5-15
	250	1	2	2	2-15 6-15, L6-15
	277	1	2	2	7-15, L7-15
	125/250	1	3	1-1/2 L-L 1/2 L-N	14-15
	250	3	3	3	11-15, L11-15 15-15
	120/208	3	4	3	18-15
20	125	1	2	1	5-20, L5-20
	250	1	2	2	2-20, L2-20 6-20, L6-20
	277	1	2	2	7-20, L7-20
	480	1	2	5	L8-20
	125/250	1	3	2 L-L 1 L-N	10-20, L10-20 14-20, L14-20
	250	3	3	5	11-20, L11-20 15-20, L15-20
20	480	3	3	10	L12-20, L16-20
	120/208	3	4	3	18-20, L18-20 L21-20
	277/480	3	4	10	L19-20, L22-20
30	125	1	2	2	5-30, L5-30
	250	1	2	3	2-30 6-30, L6-30
	277	1	2	3	7-30, L7-30
	480	1	2	7-1/2	L8-30
	125/250	1	3	3 L-L 2 L-N	10-30, L10-30 14-30, L14-30
	250	3	3	7-1/2	11-30, L11-30 15-30, L15-30
	480	3	3	15	L12-30, L16-30
	120/208	3	4	5	18-30, L18-30 L21-30
	277/480	3	4	15	L19-30, L22-30
50	125	1	2	3	5-50
	250	1	2	7-1/2	6-50
	277	1	2	7-1/2	7-50
	125/250	1	3	5 L-L 3 L-N	10-50 14-50
	250	3	3	10	11-50, 15-50
	120/208	3	4	10	18-50
60	125/250	1	3	7-1/2 L-L 3 L-N	14-60
	250	3	3	15	15-60
	120/208	3	4	15	18-60

L-L: Motor connected line-to-line.
L-N: Motor connected line-to-neutral.

Figure 1 Table of assigned horsepower ratings for attachment plugs and receptacles reproduced from the 1990 edition of Underwriters Laboratories "General Information Directory" (the so-called White Book). These horsepower ratings apply to *all*—newly manufactured and existing—listed attachment plugs and receptacles with the indicated NEMA designation numbers.

is 1hp. And, when the load is supplied at 250VAC, then the horsepower rating is 2hp. According to NEC Table 430-148, which gives the full-load current for single-phase AC motors, at 125 VAC, a 1hp motor will draw approximately 16A; and at 250VAC, a 2hp motor will draw about 12A (See Fig. 2, NEC Table 430-148). But in either case, the assigned horsepower rating is the *maximum* horsepower rating that will draw an amount of current that is approximately 80% or less of the device's ampere-rating. [Although the 12A full-load current drawn by a 250VAC single-phase AC motor is actually only 60% of the ampere-rating of a 20A attachment plug or receptacle, assignment of the next higher horsepower rating (i.e., 3hp) would result in a full-load current of 17A, which is greater than 80% of the plug's and receptacle's ampere rating (20A × 0.8 = 16A).]

This Table of ratings and accompanying statements have been added to the UL Listing Information for the categories of "Attachment Plugs, Fuseless (AXUT)" and "Attachment Plug Receptacles And Plugs (RTRT)" in the 1990 edition of the UL *Electrical Construction Materials Directory* (the so-called "Green Book") and the UL *General Information Directory* (the "White Book").

Table 430-148. Full-Load Currents in Amperes—Single-Phase Alternating-Current Motors The following values of full-load currents are for motors running at usual speeds and motors with normal torque characteristics. Motors built for especially low speeds or high torques may have higher full-load currents, and multispeed motors will have full-load current varying with speed, in which case the nameplate current ratings shall be used.

The voltages listed are rated motor voltages. The currents listed shall be permitted for system voltage ranges of 110 to 120 and 220 to 240.

HP	115V	200V	208V	230V
1/6	4.4	2.5	2.4	2.2
1/4	5.8	3.3	3.2	2.9
1/3	7.2	4.1	4.0	3.6
1/2	9.8	5.6	5.4	4.9
3/4	13.8	7.9	7.6	6.9
1	16	9.2	8.8	8
1½	20	11.5	11	10
2	24	13.8	13.2	12
3	34	19.6	18.7	17
5	56	32.2	30.8	28
7½	80	46	44	40
10	100	57.5	55	50

Figure 2 For a single-phase motor, the various devices covered by the table shown in Fig. 1 have ampere ratings that are at least 125% of the full-load current shown here for a given horsepower value. Devices for use with 3-phase motors have identically proportional ampere ratings based on NEC Table 430-150.

Application

The listing information given in the White Book spells out the types of devices in each category that are covered by the table shown in Fig. 1.

Devices covered under "Attachment Plugs, Fuseless (AXUT)" are attachment plugs and bodies, separable and non-separable attachment plugs, appliance plugs, cord connectors and bodies, including three-way taps and table taps, and motor attachment caps. Under "Attachment Plug Receptacles And Plugs (RTRT)," general use receptacles, receptacles in combination with pilot lights for use in wiring systems recognized by the NEC, and receptacles for use in appliances and fixtures are covered. And, as stated in both categories, "devices of configurations other than those indicated in the table (Fig. 1) have horsepower ratings only if such ratings are marked on the devices."

It is important to note that this table does **not** cover attachment plugs provided with integral overcurrent protection, receptacles and attachment plugs that incorporate switches, or cord connectors on manufactured cord sets. Such devices do not have a "horsepower rating," unless specifically marked, even though the blade configuration may be identical to one of those given in the table.

Additionally, it is worth noting that factory-molded or factory-assembled attachment plugs provided with equipment by the manufacturer **are not** covered by the UL standard "Attachment Plugs, Fuseless (AXUT)" and, therefore are not covered by the Table shown in Fig. 1. However, when an attachment plug is provided by the manufacturer with motor-operated equipment and when such a plug would also be permitted by the NEC to be used as the disconnecting means, the plug would be evaluated for use as the disconnect by UL when the equipment itself is evaluated. This is done even if the motor-operated machine or equipment is also equipped with another Code-recognized disconnect. The UL position is that if the attachment plug is permitted by the NEC to be used as the required disconnect, then evaluation of the completed piece of equipment must also include evaluation of the plug as the disconnecting means. Because the NEC permits use of the plug and receptacle as the disconnecting means on any "cord-and-plug connected motor," all motor-operated machines that are factory-equipped with an attachment plug by the manufacturer have already been evaluated and the plug is suitable for use as the Code-recognized disconnect.

As it now stands, even though a "horsepower rating" is not marked on a specific device, if it is one of those devices covered by the table shown in Fig. 1, the device has a "horsepower rating" as indicated in the table. And such attachment plugs and receptacles may be used as the Code-recognized disconnecting means in accordance with Exception No. 5 to Sec. 430-109. If the device is *not* specifically horsepower-

Sec. 430-109, Ex. No. 5 was revised in the 1981 and 1987 editions of the NEC to permit the use of horsepower-rated attachment plugs and receptacles to serve as the disconnecting means for all—portable or permanently installed—cord-and-plug connected motors.

rated and is *not* one of those covered by the Table or has *not* been otherwise evaluated (i.e., such as when submitted with a piece of equipment), then such a device may **not** be used as the disconnecting means. And another Code-recognized disconnect must be provided.

Some typical examples of the devices covered by the new tables in the UL "General Information Directory"—the so-called "White Book." Code-recognized horsepower ratings are given for straight-blade attachment plugs and receptacles rated from 15 to 60A, 277VAC or less, and locking-type plugs and receptacles rated from 15 through 30A, 480VAC or less.

Where a plug and receptacle is used as the disconnect for "combination loads," field-installed devices must be selected as covered by Sec. 430-110. This section requires the calculation of single-motor-equivalent full-load and locked-rotor currents to determine the minimum, acceptable rating for the devices.

Prior to the 1981 NEC, the disconnect at left (arrow) was required because the permission to use a plug and receptacle as the Code-recognized disconnecting means applied only to "portable motors." Now, if the devices are one of the types covered by the new UL tables or are specifically marked with a horsepower rating, the attachment plug and receptacle would satisfy the rule of Sec. 430-109, Ex. No. 5, and they could serve as the disconnect for this motor-operated equipment.

Certain devices, such as GFCI receptacles, are not covered by the UL table and do not have horsepower ratings, unless specifically marked, even though the blade configuration may be identical to one of those given in the UL table. Other devices not covered include plugs provided with integral overcurrent protection, receptacles and plugs that incorporate switches, and cord connectors on manufactured cord sets.

Ampere rating and interrupting capacity

NEC Sec. 430-109 describes the type of equipment that may be used as the required disconnecting means. And, as discussed above, an attachment plug and receptacle used as the disconnect must be horsepower-rated. Even though a plug and receptacle are not marked with a specific horsepower rating, the ratings indicated in the Table shown in Fig. 1 *are* the UL-recognized horsepower ratings for devices of the specific configurations shown.

It is also worth noting that the second sentence of Sec. 430-109, Exception No. 5 excludes certain applications. When an attachment plug and receptacle are used as the Code-recognized disconnect with cord-and-plug connected appliances installed in accordance with Sec. 422-22 and room air conditioners covered by Sec. 440-63, the devices do not have to be horsepower rated. Additionally, "portable" motors rated 1/3hp or less are not required to use horsepower-rated devices when the plug and receptacle are to serve as the Code-recognized disconnecting means.

While Sec. 430-109 identifies the *type* of disconnecting means, Sec. 430-110 regulates the minimum acceptable ampere rating and interrupting capacity of the disconnecting means.

When cord-and-plug connected motor-operated equipment is fitted with an attachment plug by the manufacturer, the exercise described below will not apply. For such applications, acceptability of the use of the plug as the disconnect has already been completed. And because only a similarly-rated receptacle will have a corresponding blade configuration, no further consideration is needed.

For other applications where the machine or equipment is fitted with a field-installed cord or an extension cord is made-up in the field, selected devices must comply with the rules of Sec. 430-110, in addition to being horsepower rated. If the selected devices do not also satisfy the rules of Sec. 430-110, then they may not serve as the disconnect and another Code-recognized disconnecting means must be provided.

Individual cord-and-plug connected motor

For an individual motor (with no other loads), the general requirement for the disconnecting means to have an ampere rating at least equal to 115% of the motor full-load current is no problem. As stated above, the assigned horsepower ratings are such that the full-load current for a motor of that specific horsepower rating will be 80% or less of the plug's and receptacle's ampere-rating. Another way to express this relationship is to say that the plug and receptacle will have an ampere rating that is 125% or greater of the motor's full-load current. Therefore, when using the assigned horsepower ratings to evaluate the use of a plug and receptacle as the disconnect for an individual motor load, the device will always have an ampere rating equal to "at least 115 percent of the full-load current rating of the motor."

The "Interrupting Capacity" mentioned in the title of Sec. 430-110 refers to the disconnect's ability to open the circuit under locked-rotor conditions. It is not related to short-circuit duty or withstand capability.

Generally speaking, all horsepower-rated switches, circuit breakers, and molded case switches are all tested by UL for "interrupting capacity," which is generally taken to be six times the full-load current shown in tables 430-147 through 430-150 for a motor of a given horsepower-rating. Devices of the various configurations covered in the table of Fig. 1 were subjected to a similar evaluation of "interrupting capacity." So, when these attachment plugs and receptacles are used as the disconnect for an individual cord-and-plug connected motor (with no other loads), the interrupting capacity has already been evaluated and need not be considered.

For an individual cord-and-plug connected motor (with no other loads), simply assure that the horsepower rating for the plug and receptacle is equal to or greater than that of the motor.

Combination loads

When horsepower-rated attachment plugs and receptacles are used with "combination loads"—that is, two or more motors or one or more motors with other loads in a single cord-and-plug connected machine—a "single-motor-equivalent" for full-load and locked-rotor current must be calculated to determine the minimum acceptable ampere rating and interrupting capacity [Sec. 430-110(c)].

To determine the minimum ampere rating, the full-load current value for the corresponding horsepower rating of each motor must be taken from either Table 430-148 (for single-phase motors) or Table 430-150 (for 3-phase motors). These current values are then added to the current drawn by all other loads under full-load conditions, including resistance loads. This total will be the single-motor-equivalent *full-load current* value.

To determine the minimum acceptable interrupting capacity, the locked-rotor current value for the corresponding horsepower rating of each motor must be taken from Table 430-151. Again, these current values are added to the current drawn by all other loads under full-load conditions, including resistance loads. This will be the single-motor-equivalent *locked-rotor current.*

Using this single-motor-equivalent *locked-rotor current*, refer to Table 430-151 (Fig. 3) and determine the horsepower rating that has a locked-rotor current not less than the calculated single-motor-equivalent locked-rotor current. Next, from the Table shown in Fig. 1, select a plug and receptacle that have a horsepower rating at least equal to that horsepower rating indicated in Table 430-151. Then verify that ampere rating of the selected devices is at least equal to 115% of the "single-motor-equivalent" *full-load current.*

This evaluation may seem a bit redundant because the devices covered by the Table shown in Fig. 1 have an ampere rating that is 125% or greater than the full-load current of their corresponding horsepower rating. However, the ampere rating for the selected devices must be not less than 115% of the *calculated* single-motor-equivalent full-load current, *not* 115% of the full-load current value for the corresponding horsepower rating given in Table 430-148 or Table 430-150. Always check to see that the attachment plug and receptacle comply with both requirements—interrupting capacity *and* ampere rating—when using such devices as the disconnect for cord-and-plug connected combination motor loads. This determination is essential to proper application.

Table 430-151. Conversion Table of Locked-Rotor Currents for Selection of Disconnecting Means and Controllers as Determined from Horsepower and Voltage Rating

For use only with Sections 430-110, 440-12 and 440-41.

Motor Locked-Rotor Current Amperes*							
Single Phase		Two or Three Phase					Max. HP
115V	230V	115V	200V	230V	460V	575V	Rating
58.8	29.4	24	18.8	12	6	4.8	½
82.8	41.4	33.6	19.3	16.8	8.4	6.6	¾
96	48	43.2	24.8	21.6	10.8	8.4	1
120	60	62	35.9	31.2	15.6	12.6	1½
144	72	81	46.9	40.8	20.4	16.2	2
204	102	—	66	58	26.8	23.4	3
336	168	—	105	91	45.6	36.6	5
480	240	—	152	132	66	54	7½
600	300	—	193	168	84	66	10
—	—	—	290	252	126	102	15
—	—	—	373	324	162	132	20
—	—	—	469	408	204	162	25
—	—	—	552	480	240	192	30
—	—	—	718	624	312	246	40
—	—	—	897	780	390	312	50
—	—	—	1063	924	462	372	60
—	—	—	1325	1152	576	462	75
—	—	—	1711	1488	744	594	100
—	—	—	2153	1872	936	750	125
—	—	—	2484	2160	1080	864	150
—	—	—	3312	2880	1440	1152	200

* These values of motor locked-rotor current are approximately six times the full-load current values given in Tables 430-148 and 430-150.

Figure 3 When an attachment plug and receptacle are used as the disconnect for a combination load in accordance with Sec. 430-110(c), the horsepower rating of the devices must not be less than the horsepower value that corresponds to the calculated single-motor-equivalent locked-rotor current.

Warning on Portable Droplights!

Sec. 511-3(f). WARNING! Lawsuits raise serious concerns over use of portable droplights (work lights) in auto service stations.

Over recent years, several court cases have focused on the use of portable plug-in work lights that were alleged to be the cause of fire ignition of gasoline, resulting in the injury and death by fire of auto mechanics who were working on leaking fuel systems of autos that were up on lifts. In two cases, the court found the droplight to be the cause of ignition of gasoline and made million dollar awards to the plaintiffs.

Although the designers and installers of the garage electrical system were not directly held responsible—because the work lights are portable plug-in assemblies that are not supplied as part of the fixed electrical system in the building—the entire matter of portable work lights in auto service stations is very specifically covered in the National Electric Code and could easily involve electrical design and installation personnel. All electrical design and installation personnel who work on electrical systems for auto service stations should be aware of the applicable Code rule and its relation to the lighting assemblies used in all commercial repair garages. And they should pass along a written warning to the owner or operator of every service station about the use of such lights. By submitting a warning on their company's letterhead—setting out the simple requirements of Sec. 511-3(f) of the NEC, electrical personnel can properly discharge their responsibility as electrical experts to their customers and thereby provide themselves with a documented record of having done all in their power to assure compliance with NE Code rules.

Section 511-3(f) is entitled "Portable Lighting Equipment" and is talking about the kind of trouble light (also called "work light") that is so commonly used for up-close lighting of various locations where gas-powered vehicles require servicing—both in the under-hood engine compartment and underneath a vehicle that is up on a service lift. The most commonly used droplight is the usual cord with a plug-cap at one end and a lamp assembly, with shield, at the other end. This is the usual assembly that everyone is familiar with—the type available at any hardware store or home center. But that type of assembly falls far short of meeting the requirements of Sec. 511-3(f) and is **not** at all acceptable for use in an auto repair garage, because the Code calls for the following:

1. The portable lighting unit must be equipped with a handle, a lampholder, a hook for hanging, and a "substantial guard" attached to the lampholder or handle. On that score, they are OK.
2. All exterior surfaces must be of nonconducting material or covered with insulation. And that's readily satisfied. But WATCH OUT for the next rule!
3. The lampholder must be "of the unswitched type" (that is, no push or rotary switch in the lampholder assembly), and there must **not** be a receptacle in the lampholder assembly. The usual work light has both a switch and a receptacle as part of the lamp-head assembly—which makes it a violation to use such a work light in an auto repair garage.
4. Then Sec. 511-3(f) says that any portable lighting equipment must be "approved for Class I, Division 1 locations," UNLESS the lamp and its cord are "supported or arranged in such a manner that they cannot be used" in the Class I, Division 2 location that extends up to 18 inches above the floor of the garage. That is, the portable lighting assembly does **not** have to be UL-listed for a Class I, Division 1 location **if** the lamp-head assembly is physically restrained from entering the 18-in. space under any possible conditions.

Retractable-reel type work lights are assemblies with the lamphead on the end of a cord that can be mechanically retracted (wound up) into an overhead reel housing. This is shown in Figure 1 at the arrows up at the steel joists of the service area. BUT the length of cord and restraint on the cord extension from the reel must be such that if the cord is pulled straight down, the lamp-head will be more than 18 in. above the floor. As shown in Figure 2, those cord and lamp-heads can be, and usually are, arranged so the lamp-head could readily be in the Class I, Division 2 area—even sitting right on the floor. Because such portable reel-type portable light units do not meet the requirements of Sec. 511-3(f), it is a Code violation to use them in an auto garage.

All that analysis presents a truly perplexing situation. The regular type of pull-cord portable work lights **cannot ever** be used in such garages. The reel units can be used if they actually provide a positive restraint from entering the hazardous 18-in.-high space. Otherwise, a work light in a repair garage must be listed and labeled for Class I, Division 1 hazardous locations. That's a tough one.

Figure 1 Lamp-head (arrow) of work light assembly is on end of cord, which can be pulled off the overhead reel that stores the cord in its retracted coil. This is OK if lamp-head cannot be pulled straight down into the hazardous space that extends up to 18 in. above the floor.

Figure 2 The extended length of the cord, pulled out of an overhead reel-type lighting assembly, is so long that the lamp-head could easily be placed in the classified (hazardous) space up to 18 in. above the floor.

Wiring for Computers

Article 645. In the 1990 NEC, a major effort was put forth by Code Making Panel (CMP) 12 to rewrite the rules of Article 645. The primary thrust of this effort was to better define exactly what is and what isn't a computer room; illuminate the various requirements for wiring within such spaces; incorporate clarified and additional requirements for the emergency disconnecting means; and establish the grounding requirements for such equipment and systems. The many complex and confusing requirements for computer installations contained in the radically expanded and revised rules on that subject in the 1990 NEC must be carefully correlated to assure safe practice in day-to-day design and installation work.

Penetrating the practical meaning of these new NEC rules for computer rooms requires careful attention to not only the rules of Article 645, but also other existing and new related Code sections. The following discussion will focus on the major changes and additions to Article 645 and the other related rules to provide some understanding as to what they say and what should be done to assure a safe, functional, and Code-complying installation.

What is a "computer room"?

The first part of this Article that will be examined is Sec. 645-2. This section gives six criteria that are to be used for determining whether or not Article 645 applies to a particular installation—that is, if a particular installation is or is not a computer room as far as the NEC is concerned. If these six conditions do **not** exist, then this Article does **not** apply, and the room and the equipment within the room must be wired in accordance with the requirements given in Chapters 1 through 4. (To help facilitate this discussion, Sec. 645-2 has been reproduced in its entirety in Fig. 1.)

As given under part 1 of Sec. 645-2, a disconnecting means that complies with the requirements of Sec. 645-10 must be provided. It seems a bit odd to make compliance with this Code article dependent upon compliance with one of the rules given within the article; but, if a disconnect is not provided or if one that does not completely satisfy the requirements of Sec. 645-10 is provided, then the permission and requirements of Article 645 do not have to be followed. However, any

Is this a computer room? That depends upon whether or not the six conditions described in Sec. 645-2 have been met. If not, then, even though there are a great many computer terminals within these spaces, such a location is *not* an "electronic computer/data processing room."

75-1989 (ANSI).

645-2. Special Requirements for Electronic Computer/Data Processing Equipment Room. This article applies provided all the following conditions are met:

(1) Disconnecting means complying with Section 645-10 are provided.

(2) A separate heating/ventilating/air conditioning (HVAC) system is provided that is dedicated for electronic computer/data processing equipment use and is separated from other areas of occupancy. Any HVAC system that serves other occupancies may also serve the electronic computer/data processing equipment room if fire/smoke dampers are provided at the point of penetration of the room boundary. Such dampers shall operate on activation of smoke detectors and also by operation of the disconnecting means required by Section 645-10.

(FPN): For further information, see Protection of Electronic Computer/Data Processing Equipment, NFPA 75-1989 (ANSI).

(3) Listed electronic computer/data processing equipment is installed.

(FPN): For further information, see Protection of Electronic Computer/Data Processing Equipment, NFPA 75-1989 (ANSI), Section 4-1.3.

(4) Occupied only by those personnel needed for the maintenance and functional operation of the installed electronic computer/data processing equipment.

(FPN): The computer room is not to be used for the storage of combustibles beyond that necessary for the day-to-day operation of the equipment. For further information, see Protection of Electronic Computer/Data Processing Equipment, NFPA 75-1989 (ANSI).

(5) The room is separated from other occupancies by fire-resistant-rated walls, floors, and ceilings with protected openings.

(6) The building construction, rooms, or areas and occupancy comply with the applicable building code.

Article 645 i[...] equipment and wiring locat[...] the data proc[...]

Figure 1 These are the six "pre-conditions" that determine whether or not the rules and permissions of Article 645 apply. If *all* of these conditions exist, then Article 645 *does* apply to the installation of equipment and conductors. If *all* of these conditions *do not* exist, then only the general rules of Chapters 1 through 4 of the NEC must be observed.

portion of Article 645 may be arbitrarily applied provided that it does not violate another rule given in Chapters 1 through 4. Nevertheless, whether or not your installation complies with Article 645, a prudent and safe position would be to install this disconnect. It could save lives.

The second requirement in Sec. 645-2 is not entirely clear. It appears to call for either a dedicated HVAC system for the equipment *and* room, or a dedicated HVAC system for the "*equipment*" and use of "any" HVAC, which may serve other spaces, to condition the "*room,*" but only if the HVAC system is properly dampered to shut down the return and supply air upon operation of the emergency disconnects (Sec. 645-10) or the "smoke detectors" [See underlined portions of Sec. 645-2(2)]. By specifically requiring the "equipment" to be supplied by a "separate...HVAC system that is **dedicated** for...equipment use **and separated** from *other* areas of occupancy," and then providing no other rule for "equipment" HVAC, it can only be concluded that there must be a dedicated HVAC system for the equipment. The net effect of such a rule would be to prohibit anything other than a dedicated air conditioner for supplying the raised floor area, which is normally used for equipment cooling. This conclusion is corroborated by Sec. 645-5(d)(3) where the Code permits the use of various power and data/comm conductors, cables, receptacles, etc. under the raised floor IF

> "Ventilation in the underfloor area is used for the data processing equipment and data processing area only."

Although the term "principal exit doors," as referred to in Sec. 645-10 is not completely clear, if a means for controlling the disconnect required by this section is *not* located at *every* exit door to an electronic computer/data processing room, then a means for such control should, at least, be located at every *marked* exit door.

As now required by new Sec. 645-5(d)(4) openings in the raised floor where cables emerge must provide protection against abrasion and "minimize entrance of debris." An installation such as this would be prohibited.

Load-side conductors of computer power equipment are generally considered to be "premises wiring." However, if the computer power equipment is listed in accordance with UL 478, then the load-side conductors might not be considered to be premises wiring. And such conductors would *not* have to be fastened in place as called for by Sec. 300-11.

In addition to clarifying the concept of a dedicated equipment HVAC system, this section helps point up the fact that the dedicated equipment-system may also be used to condition the data processing room, which also seems to be permitted by the wording of the second sentence of Sec. 645-2(2). By saying "Any HVAC system that serves other occupancies may also serve the...room," it would be reasonable to say that the dedicated equipment-system qualifies as "any HVAC system." And this system is *required* to supply *only equipment* through the underfloor space. It would also be reasonable to categorize the equipment HVAC system as one that "also serves other" areas because the first sentence of Sec. 645-2(2) and Sec. 645-5(d)(3) effectively require the underfloor space to be treated as a separate area.

Because a dedicated HVAC system must be provided for the equipment, consideration should be given to increasing the size (BTU capacity) of this air conditioning unit so that it can provide conditioning of the room as well. This will eliminate the Code-requirement for smoke detectors, dampers, their associated controls, and additional circuiting to the emergency disconnects required by Sec. 645-10. (Although the NEC would permit eliminating smoke detectors, good practice and most building codes make their use mandatory.)

Another concern that is not addressed here, or elsewhere in Article 645, is the operation of the smoke detectors called for when an HVAC system that would require dampering is used to cool the computer room. In terms of operation, would it be permitted to generate an alarm and system shutdown on activation of at least two smoke detectors in two zones, or must the alarm and shutdown occur after activation of a single detector in a single zone? Certainly, if the latter approach is used, data centers will be subject to repeated, unnecessary, and potentially very costly, nuisance shutdowns, which from an operational standpoint are extremely objectionable. Because the Code does not specifically address this consideration, use of a two-detector/two-zone alarm and shutdown scheme (hopefully) should be acceptable.

Additionally, questions exist about the use of sprinklers. For instance, would use of an overhead sprinkler system and smoke detectors beneath the raised floor only satisfy the requirement given in this section? If so, should activation of the sprinkler system also cause an alarm and shutdown? No mention is made of sprinklering.

There are many fire protection engineers who believe sprinklering is the way to go; others disagree and say that detection is better. If there is disagreement as to the more effective method, both methods—sprinklering and smoke detectors—should be recognized. Due to lack of specific guidance in the NEC on these matters, the best (and only) course of action is to check with the local inspection authority to determine what will be acceptable.

The next requirement is Sec. 645-2(3). The rule states that only "listed" equipment is to be used. While a requirement to use only equipment that has been tested and found to be essentially safe when used in accordance with its listing instructions is in itself a good rule, it is not practical and completely beyond the control of the designer or contractor. If, for example, an installation were completed in accordance with Article 645, how is the engineer or installer to prevent the use of a non-listed piece of equipment after the space is occupied? Also, how much of the new imported data equipment is actually listed by UL, ETL, Factory Mutual, or some other independent nationally recognized testing lab?

The next requirement is also beyond the control of the designer or installer. Sec. 645-2(4) requires that the computer room be "occupied only by those personnel needed for the maintenance and functional operation of the...equipment." Most data centers are secure areas and security personnel would not likely be considered as necessary for the "maintenance and functional operation" of the computer equipment. This rule would effectively require that security officers, administrative personnel, secretaries, etc., be stationed or located outside the data center. Again, this cannot be controlled by the engineer or contractor, and the Code Making Panel (CMP) should not use criteria such as this, and the requirement of Sec. 645-2(3), to differentiate between what is and isn't a computer room, and whether Article 645 does or doesn't apply. Nonetheless, be advised that use of non-listed equipment and/or occupancy by non-essential personnel would violate the requirement(s) of Sec. 645-2(3) and/or Sec. 645-2(4), respectively, and the installation would not be covered by Art. 645 but would have to comply with the requirements of Chapters 1 through 4.

Sec. 645-2(5) specifies the type of construction required for walls, floors, and ceilings in the computer room and Sec. 645-2(6) calls for "the building construction, room, or areas and occupancy" to comply with the local building codes. The fundamental problem with this Code rule is that it is not written to the ultimate approver—the electrical inspector. Generally, an electrical inspector is not qualified with respect to *all* local building codes. In addition, the Code gives the inspector (or designer, or contractor, for that matter) no guidance as to what the minimum acceptable fire-resistance rating should be. Undoubtedly, the result will be discrepancies between what is required from one jurisdiction to another. Again, the safest bet is to consult the local inspecting agency for *their* requirements to assure acceptance of the finished project.

It is most unfortunate that the CMP approached the task of defining an electronic computer/data processing center from a non-assertive or non-mandatory position. Instead of saying that the rules of Article 645 apply *if* these requirements are met, this section should be made mandatory and *all* data processing centers should be required to comply with the Code-prescribed criteria.

As it now stands, the distinguishing of a work center, a communications room, etc., from a data center is somewhat confusing, subject to interpretation; and, as a result, not completely clear. What is clear, though, is the fact that the designer or installer can arbitrarily invoke or eliminate the requirements of Article 645 simply by *not* complying with one of the six preconditions.

Wiring methods

Sec. 645-3 in the 1987 NEC has been revised and expanded and now appears as Sec. 645-5 in the 1990 edition. Here the NEC gives requirements for branch-circuits; connecting cables, which are the connections between the power system outlet and the electronic computer/data processing equipment; interconnecting cables, which are connections between electronic computer/data processing equipment and/or peripherals; and for wiring in the underfloor area.

The rules of Sec. 645-5 remain essentially the same as they were in the 1987 edition of the NEC with the exception of some editorial changes. One shortcoming here is the failure to specifically require only low-smoke producing, fire-resistant insulations for all the connecting and interconnecting cables when they are installed beneath the raised floor.

Certainly, the underfloor area is similar in form and function to a hung-ceiling space used for the supply and/or return of environmental air. As covered by Sec. 300-22(c) for general work, the wiring methods permitted for other spaces used for environmental air is limited to cables and conductors that are "specifically listed for the use" or, when the conductors or cables are not so listed, they must be installed in one of the permitted raceways. Similar restrictions should be placed on the cables and conductors used in the underfloor area of a computer room. This becomes especially clear when considering the rule of Sec. 300-22(d), which reads:

> Electric wiring in air-handling areas beneath raised floors for data processing systems shall comply with Article 645.

Does this rule require that the space or the room first comply with the requirements of Sec. 645-2? It should, but this is not clearly spelled out. If the space or room does *not* comply with the requirements of Sec. 645-2(2), then there exists the possibility that Sec. 645-5(d)(3) will not be interpreted properly. When the requirements of Secs. 645-2(2) and Sec. 645-5(d)(3) are properly correlated, it is clear that the underfloor area is not permitted to be ventilated by anything except a dedicated HVAC; and, therefore, any toxic fumes that may result during a fire would not be circulated to adjacent occupancies of the building. But, if the room and underfloor area for a "data processing system" are not ventilated in a manner that isolates and restricts the air transfer (as described in Sec. 645-2), then there exists the possibility that toxic fumes will be circulated to other occupancies during a fire.

The only rational, logical, and safe approach to this conundrum is to apply the rules of Article 645 *only* when the preconditions of Sec. 645-2 have been completely met. Otherwise, treat the underfloor area just the same as any "other space used for environmental air" and use only those wiring methods permitted by Sec. 300-22(c).

One addition to the raceways permitted for use with branch circuits installed beneath raised floors in computer rooms is liquidtight flexible nonmetallic conduit [Sec. 645-5(d)(2)]. This addition fills a void that existed in previous editions of the Code for 415Hz applications. Prior to the 1990 NEC, only ferrous metal flex was recognized because flexible aluminum conduit is not listed due to lack of crush resistance. In 415Hz applications, the ferrous metal flex introduces excessive and sometimes unacceptable losses due to the inductive reactance of steel raceways. As a result, the use of the newly recognized flexible nonmetallic conduit will find widespread application for 415Hz computer power distribution.

For a number of years now, there has been disagreement as to whether or not conductors and cables installed beneath the raised floor were required to be secured in place as is generally required by Sec. 300-11. Sec. 645-5(d)(2) was amended and Sec. 645-5(e) was added to help clarify what conductors and cables would and what ones wouldn't have to be secured.

Sec. 645-5(d)(2) gives the types of raceways and wiring methods permitted for branch-circuit supply conductors under a raised floor. And the last sentence of this rule states that all branch-circuit supply conductors must be installed in accordance with Sec. 300-11, where the Code requires all "raceways, cable assemblies, boxes, cabinets, and fittings shall be securely fastened in place." Notice that the wording of Sec. 300-11 calls for the cables and equipment to be "fastened in place" and does not simply require that they be "supported." Therefore, even though the cables and equipment are supported by the floor, they still have to be "securely fastened" in place—such as to the raised-floor pedestals—at the appropriate intervals called for in the individual Code article that covers the wiring method used.

In the new section, Sec. 645-5(e), however, "power cables, communications cables, connecting cables, interconnecting cables, and associated boxes, plugs, and receptacles that are listed as part of, or for, electronic computer/data processing equipment" are exempt from this requirement.

Because communications cables are not subject to the general requirements given in Chapters 1 through 4, and because Article 800 has no requirements for "securely fastening" such cables in place, communication cables are not, and never have been, required to be secured. Additionally, "connecting cables," their associated plugs and boxes, and "interconnecting cables" *must* be "listed" for the use, as called for in Secs. 645-5(b) and (c); and, therefore, such equipment is also exempt from the requirement of Sec. 300-11. But what about "power cables"?

Use of the word "cables" presents an immediate problem. Is it the intent of this Code-rule to limit permission to only "cables," or does it also extend to factory-assembled flexible whips? The rational answer should be "yes," but the Code is not clear.

Generally speaking, all branch-circuit conductors, even those conductors extending from the load side of an uninterruptible power supply (UPS), a computer power unit (CPU) or a power distribution power unit (PDU), must be secured UNLESS the CPU or PDU is listed in accordance with UL 478 "Data Processing Equipment, Electronic." While, to the best of my knowledge, there are no UPS systems so listed, if the CPU or PDU *is* so listed, then the load side conductors would *not* be considered "branch circuit" conductors. Instead, the load side conductors would be considered to be "interconnecting cables" and therefore, they would *not* have to be secured. In addition, they may be just cables installed without raceway.

Disconnect and UPS

In the 1987 NEC, the emergency disconnecting means required by this Article was defined in Sec. 645-3. These requirements have expanded slightly and now appear in Sec. 645-10.

Generally, this rule calls for a disconnecting means to be provided that will deenergize all power to the electronic equipment in the data processing room, the HVAC system(s), *and* operate the fire/smoke dampers (if used). This may be accomplished by using a single "means" or the disconnecting functions may be initiated by individual "means"—a CB, fused switch, or remote shunt-trip actuator for each.

The '90 Code now also requires the disconnect means to be "grouped and identified." That is, the emergency disconnects must be durably and legibly marked and installed in close proximity to each other in one location. And, as the wording of the rule states: "shall be *controlled* from locations readily accessible at the principal exit doors." If a computer room only has a single door, the wording of this rule presents no problem. However, if there is more than one door, then things get a bit more involved.

The wording of this rule refers to "the *principal* exit *doors*" and no definition or other guidance is provided as to what is meant by that. Additionally, the use of the plural "doors" indicates that the CMP is of the opinion that there may be more than one. How does one decide which door is a "principal" exit and which is not? Short of consultation with the local inspecting authority, there is no way for the designer or installer to know. While a better feeling for what is meant and required by this rule will be developed over time, the easiest approach, and one that is guaranteed to always be acceptable, is to provide for control of the required disconnecting means at *all* doors or at least those marked "EXIT."

Another completely new requirement is given in Sec. 645-11. This rule calls for UPSs installed within the data center to also be isolated—input and output circuits opened—upon activation of the disconnecting means required by Sec. 645-10; and the last sentence requires that the disconnecting means also "disconnect the battery from its load."

This wording appears to concern itself with static UPS systems only. All static UPSs have a rectifier section, an inverter section, and a battery section. The only performance difference is whether the UPS is the "off-line" or "on-line" type. But, when it comes to rotary UPS systems, these are varied in design with many showing quite an amount of imagination. Many do not have batteries. One rotary system, for example, is driven by a diesel engine. Is it the intent of the last sentence of Sec. 645-11 to require that other stored or backup energy sources, besides just batteries, also be isolated? This is not clear.

Another consideration deals with the wording here that requires the battery to also be disconnected from "its load." In a static UPS, the battery is connected through the DC bus to the inverter. The inverter output is required by the first sentence of Sec. 645-11 to be opened upon activation of the emergency disconnect. When this occurs, the battery is only feeding the inverter. Does the CMP consider the inverter to be a "load"? If not, then opening of the input and output circuits of the UPS, only, would satisfy the literal wording of this rule. If so, then the battery must also be disconnected from the DC bus to comply with this rule.

While the Code is not clear on these points, the best approach would be to always assure that the input, the output, and the battery or other stored or backup energy sources are all disconnected or isolated. In most applications, this can be readily accomplished by specifying or requiring that remote shunt-trip capability for the emergency cut-out switch be provided. (The emergency cut-out switch is standard on most UPS systems and opens the input, output, and battery CBs. The remote-trip capability is an available option on the vast majority of UPS systems, and will provide for remote operation of the emergency cut-out switch.) For the more exotic rotary systems that do not employ batteries as the short-term alternate power source, other arrangements must be considered to assure that the UPS is completely isolated to stop the power flow in an emergency.

Grounding

The most serious, and potentially dangerous, revision is in Sec. 645-15, which covers grounding for electronic computer/data processing equipment and computer power systems derived "within listed electronic computer/data processing equipment." This particular topic has been the source of much confusion and controversy for a number of Code editions.

In the 1987 edition of the Code, the requirements for grounding, as then covered in Sec. 645-4, were extremely brief and to the point. It read:

> "All non-current-carrying metal parts of a data processing system shall be grounded in accordance with Article 250."

While this would appear to make clear that the *safety* grounding generally required by Article 250 must be satisfied, there were a couple of problems reconciling this requirement with certain other Code rules.

The first had to do with a rule that has been in the Code for a number of editions: Sec. 250-21. This section permits interruption or disconnection of the required safety grounding connections to eliminate "objectionable current over grounding conductors." Those individuals (usually management information system or data processing managers) who felt that the required bonding connection was causing data errors or other computer malfunctions, often referenced this section as the Code-recognized permission to disconnect the bonding jumper between the neutral and grounding conductors. CMP 5, however, in the

THE BASICS OF "RIGHT" AND "WRONG" GROUNDING

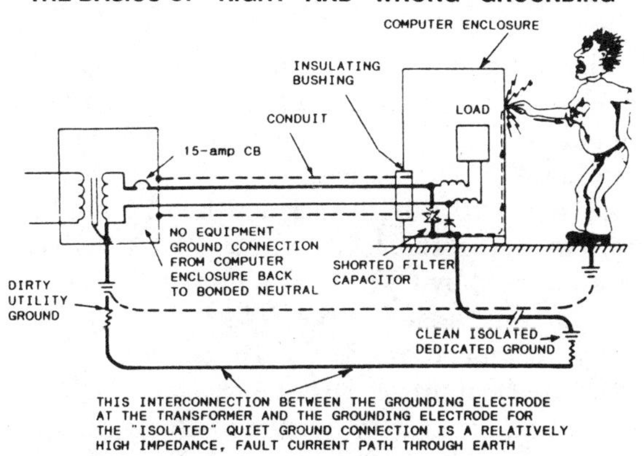

METAL CONDUIT AND EQUIPMENT GROUNDING CONDUCTOR WILL CARRY ENOUGH FAULT CURRENT TO TRIP BRANCH CB IN CASE OF FAULT AT LOAD

COMPUTER ENCLOSURE

METAL BOND OR INSULATED

EQUIPMENT GROUNDING CONDUCTOR IN CONDUIT

OPTIONAL GROUNDING CONNECTION TO ANOTHER GROUNDING ELECTRODE

Computer grounding concepts are shown in these two sketches adapted from the publication, "Guideline on Electrical Power for ADP Installations"—a Federal Information Processing Standards (so-called "fips") Publication from the U.S. Department of Standards (Institute for Computer Sciences and Technology). Known as "FIPS PUB 94" or just "FIPS 94," this 99-page book is for sale by the National Technical Information Service, U.S. Department of Commerce, Springfield, VA. 22161. The sketch at the top shows that a shock hazard will exist at any computer equipment enclosure when the necessary raceway or cable feeding the computer enclosure does not include an equipment grounding conductor or provide a continuous metallic path from the computer enclosure back to the bonded neutral point of the transformer from which the computer branch circuit is run. An insulation failure in the load equipment—such as the shorted filter capacitor indicated—can put a dangerous voltage on the metal computer enclosure which is then exposed to contact by a person who will be shocked or electrocuted, depending upon his contact with ground (concrete floor, building steel, or some other conductive path). Only 75 to 100mA of current can be lethal. And such an accident can occur even if the computer is connected to a so-called "clean, isolated dedicated ground"—such as to a separate ground rod or other separate electrode that is not connected to the transformer neutral. Without a low-impedance equipment grounding conductor (metal conduit and/or separate bare or insulated equipment grounding conductor) connecting the computer enclosure to the bonded transformer neutral, not enough fault current will flow to trip the 15A breaker, thereby quickly and automatically deenergizing the circuit and removing the hazardous voltage from the enclosure. The sketch at bottom shows how an effective hookup of safety grounding will prevent the conditions shown in the top sketch. If a fault occurs and places a voltage on the enclosure, the circuit breaker will immediately trip open, protecting personnel. In the top sketch, the use of an insulating bushing (or other nonmetallic fitting) at the connection of the metal conduit to the metal computer enclosure was a violation of the 1987 NEC and previous editions but is recognized by the Exception to Sec. 250–75 of the 1990 NEC. However, where such a nonmetallic break is placed in a metal raceway run, the new Exception requires that an insulated equipment grounding conductor must be run in the raceway *AND* the nonmetallic fitting must be listed for connection to metal raceway. That Exception to Sec. 250–75 would permit use of a nonmetallic connection of the metal conduit at the computer enclosure in the bottom sketch (at the point marked METAL BOND OR INSULATED) if the equipment grounding conductor is insulated.—Ed.

1990 NEC eliminated this argument by adopting new Sec. 250-21(d). This new section specifically prohibits interruption of the required bonding and grounding paths where this is done only to reduce or stop "currents that introduce noise or data errors."

Another argument had to do with whether or not the NEC applied to the conductors and equipment on the load side of PDUs, CPUs, or UPSs. This matter was addressed by CMP 1. The phrase "or source of a separately derived system" was added to the definition of "premises wiring" in Article 100. By all indications in the Technical Committee Report (TCR) and Technical Committee Documentation (TCD), this was intended to clarify that *all* separately derived systems, including PDUs, CPUs, and UPSs, were subject to the rules of the NEC.

These two changes are to be applauded. Safety grounding of systems and equipment is critically important and must not be compromised.

That's the good news. The bad news is the rewrite of Sec. 645-15 and the new Exception to Sec. 250-75. Let's take a look at where we stand, today.

Sec. 645-15 now reads as follows:

> 645-15. Grounding. Electronic computer/data processing equipment shall be grounded in accordance with Article 250 or double insulated. Power systems derived within listed electronic computer/data processing equipment that supply electronic computer/data processing systems through receptacles or cable assemblies supplied as part of this equipment shall not be considered separately derived for the purpose of applying Sec. 250-5(d). All exposed noncurrent-carrying metal parts of an electronic computer/data processing system shall be grounded.

The first and last sentences of this rule really present no safety problem. But the second sentence appears as if it could literally be fatal! From only a hazy indication in the TCD, it seems to be the CMP's intent that PDUs and CPUs that are listed in accordance with UL 478 as "Data Processing Equipment, Electronic" *do not* have to comply with the grounding requirements given in Sec. 250-26 for "separately derived systems." (See Fig. 2.)

This erroneous concept is further supported by the listing instructions given on page 234 in the 1990 edition of the UL publication *Electrical Appliance and Utilization Equipment*. The text, which is partially reproduced in Fig. 3, notes that equipment listed in accordance with this standard "require special installation such as...special grounding methods." This instruction further states that "such features are covered in the manufacturers' installation instruction." Strangely, if the manufacturer's instructions for any PDU, CPU, or other power system listed in accordance with UL 478 states that the neutral is *not* to be bonded and grounded, then to do so is to violate the

STEP 1 – BONDING JUMPER

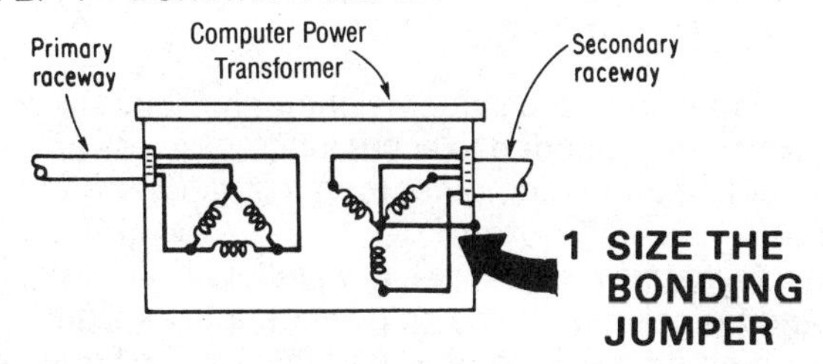

STEP 2 – GROUNDING ELECTRODE CONDUCTOR

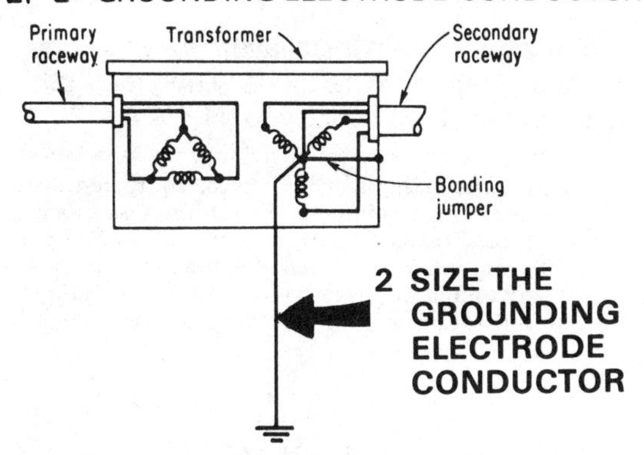

STEP 3 – GROUNDING ELECTRODE

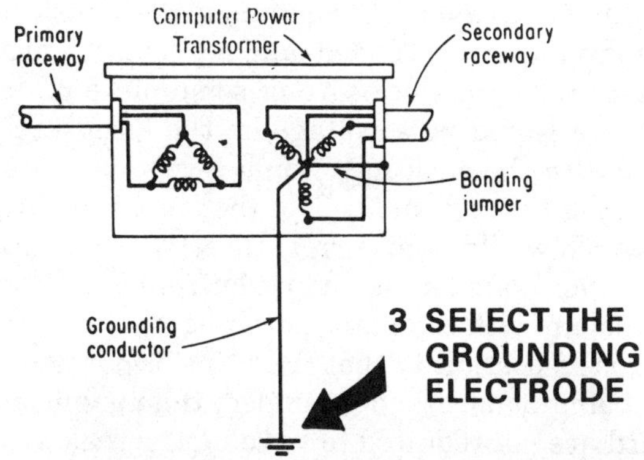

Figure 2 The rules of Sec. 250-26. To avoid accidental shock or electrocution, when given an option by the manufacturer, *always* bond and ground computer power equipment listed in accordance with UL 478 just as if it were any other "separately derived system."

DATA PROCESSING EQUIPMENT, ELECTRONIC (EMRT)

This listing covers individual units and systems primarily electronic in function and design, which are intended to accumulate, process, or store data and are intended for use in computer rooms or other areas set aside for the purpose.

Many of these units and systems require special installation such as separate transformer and branch circuit power, power supplies, special grounding methods, high frequency motor generator equipment, air conditioning, etc. Such features are covered in the manufacturers' installation instructions.

The individual units comprising a system installation are designed to be interconnected by means of one or more of the wiring methods outlined in Article 645 of the National Electrical Code (NFPA No. 70).

Figure 3 The first three paragraphs from the listing instructions for equipment listed in accordance with UL 478, "Data Processing Equipment, Electronic," as given on page 234 in the 1990 edition of the UL publication *Electrical Appliance and Utilization Equipment*. Because Sec. 110-3(b) requires that all listed equipment be used in accordance with listing instructions, any computer power equipment listed under this standard would have to be grounded in accordance with the manufacturer's installation instructions.

requirement of Sec. 110-3(b), which calls for listed equipment to be used in accordance with its listing instruction.

The problem with this is, if such a piece of equipment were installed without the neutral and ground buses bonded, and a phase conductor came in accidental contact with the enclosure of some piece of data processing equipment, there would be a voltage on the enclosure. A person coming in contact with the enclosure would receive a shock—he would have 120VAC pass through his body either hand-to-feet or hand-to-hand, either of which would send current directly through the heart. Obviously, this is a very dangerous and undesirable condition.

This is not the only problem with computer power equipment listed per UL 478. In addition to the possibility that the transformer neutral may be *prohibited* from being bonded and grounded, the panelboards listed under this standard are *not* required to have main breakers, as would normally be required by Sec. 384-16(d). Another shortcoming in stating that such equipment is not to be considered a separately derived system is that the load side conductors would not be "premises wiring" and, therefore, not subject to inspection. Additionally, they do not have to be of any specific type or insulating material.

Correlating the various NEC requirements, it becomes clear that:

1. If the room meets the six preconditions given in Sec. 645-2 *AND* the computer power equipment is listed in accordance with UL 478,

then the equipment must be grounded in accordance with the manufacturer's instruction, which may or may not prohibit bonding and grounding the secondary neutral.

AND

2. If the room *does not* meet the six preconditions given in Sec. 645-2 *OR* the computer power equipment is *not* listed in accordance with UL 478, then the equipment must be grounded in accordance with Sec. 250-26 for separately derived systems.

AND

3. When computer power equipment is listed in accordance with UL 478 *AND* is prohibited from having the neutral bonded and grounded by the manufacturer's instruction, such equipment may only be used for computer power applications where the room meets the six preconditions given in Sec. 645-2.

If the computer power equipment is listed per UL 478 and the manufacturer gives an option to bond and ground or *NOT* to bond and ground, always provide bonding and grounding. For safety's sake, ignore the permission given to operate the system ungrounded.

The best way to avoid any problems with the rule of Sec. 645-15 is to always require that such equipment be listed in accordance with UL 67 "Panelboards" and UL 506 "Speciality Transformers." When this is done, the rule of the second sentence in this section does *not* apply because the power systems are not "within listed electronic computer/data processing equipment" and safety grounding as required by Sec. 250-26 *must be* provided. Also, the other shortcomings associated with equipment listed in accordance with UL 478 are eliminated.

The permission given in the new Exception to Sec. 250-75 seems to conflict with the new Sec. 250-21(d). This Exception permits interrupting the normally required continuous electrical path in metal conduits when "required for the reduction of electromagnetic noise (electromagnetic interference)." In addition to conflicting with the new rule of Sec. 250-21(d), prohibiting altering of the grounding paths to eliminate "objectionable currents"—such as electromagnetic noise and interference—the new Exception to Sec. 250-75 is *extremely dangerous*!

When using any isolated grounding system, safety always becomes marginal. The impedance of Nos. 14, 12, and 10 copper conductors is considerably higher than the enclosing metal raceway. Especially when the magnetic path is broken. This was proven a number of years ago and reported in IEEE paper #54-36, published in 1954, by Besson and Richeau. They demonstrated through actual tests that whenever the magnetic path of an enclosing ferrous metal conduit is opened (as would be permitted by the new Exception to Sec. 250-75) the impedance of the ground-fault circuit doubles. When this is added to the conductor impedance, it becomes obvious that under line-to-ground fault conditions, lethal voltages will be present between the enclosure connected in accordance with Sec. 250-75, Exception and the raceway system, or another nearby panelboard, or, indeed, any grounded metal! Personal experience has shown the same results.

In one application, a box was surface mounted with a duplex receptacle at a distance of approximately 2 ft from a computer enclosure. After conducting a line-to-ground fault calculation (ignoring arc drop), a 60V potential difference was found between both pieces of equipment! Had the raceway system been "opened," the voltage would have been considerably higher. Always be on the safe side of this issue. Don't use the permission given in the Exception to Sec. 250-75. This is not mandatory, this is an option. If it is felt that such permission must be exercised, then consider the use of a nonmetallic raceway or, at the very least, use aluminum. DO NOT USE FERROUS METAL RACEWAYS!

It is very important to understand the Code's position on grounding and bonding. The grounding requirements for electrical equipment and systems, as given in NEC, are only intended to assure a low-impedance path for the return of fault current to actuate the circuit protective device, to thereby minimize the possibility of shock or electrocution. If a piece of equipment doesn't operate as it should when proper safety grounding as called for in Article 250 is used, there is a flaw in the equipment, *NOT* in the required grounding.

Load Pickup and Shedding Controller Requirement Clarified

Sec. 700-5(b). Load pickup and load shedding controls are *not* required when a generator has a rating equal to or greater than the amount of emergency *and* standby load to be supplied.

In a recent electrical installation for a large supermarket, the electrical inspector raised the question about the need for automatic load pickup and load shedding controls for a 45-kW generator that was installed to handle 19 kW of emergency lighting and a total of 21 kW of optional standby load (cash registers and other machines essential to store operations). Although the generator capacity was substantially in excess of the total connected load of 40 kW, the inspector was reading Sec. 700-5(b) in a strict and narrow meaning. That rule says a generator "shall be permitted to supply emergency, legally required standby, and optional standby system loads where automatic selective load pickup and load shedding is provided as needed to assure adequate power to (1) the emergency circuits; (2) the legally required standby circuits; and (3) the optional standby circuits, in that order of priority." (See Figure 1.) The inspector was reading the rule as flatly

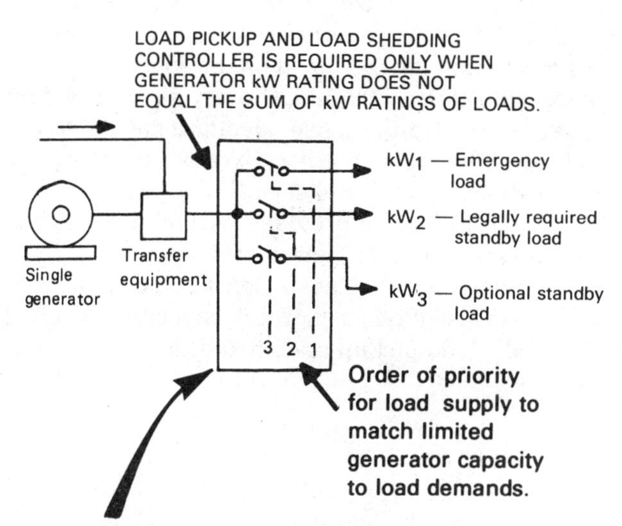

Figure 1

requiring load pickup and load shedding control in *any* installation where a single generator is used to supply both emergency *and* standby load. He was reading the above-quoted rules as if it ended with the phrase "...where automatic selective load pickup and load shedding is provided."

Reference to NFPA documentation at the time this rule was added to the NEC (the Technical Committee Reports and Technical Committee Documentation for the 1981 NEC) makes clear that the intent of the rule is to *require* the load pickup and shedding controller *only* when the generator has a kilowatt capacity less than the kilowatt sum of the emergency and standby loads. When the generator capacity exceeds the kilowatt rating of the total connected load (emergency plus standby), there is clearly no need for controls to establish the "priority" of load connection—the entire connected load can be fed at the same time.

However, when the generator capacity is *less* than the total of emergency and standby load, the Code rule calls for the load pickup and shedding controls to assure that emergency load has priority over standby load. The highest priority is that generator capacity must be committed first to whatever amount of emergency load is calling for power; and, if that is satisfied at some kilowatt rating less than the generator rating, the surplus power may be supplied to standby load. But if there is an increase in demand for power to emergency load, a corresponding amount of standby power must be shed to make the generator power available to the emergency need. Of course, as required by part (**a**) of Sec. 700-5, the generator capacity must always be at least equal to the maximum emergency demand load.

In the wording of the rule of Sec. 700-5(b), the key as to the intent of the rule is the phrase, "as needed to assure adequate power" to emergency and standby loads. That phrase critically modifies the phrase, "automatic selective load pickup and load shedding is provided." It literally says that load control must be provided *only* if such control is *needed* to prioritize commitment of generator capacity—which is needed only when a generator rating is not at least equal to the sum of emergency and standby loads. When generator capacity exceeds the sum of all connected loads, load pickup and shedding are *not* "needed to assure adequate power."

Marking Emergency Circuit Boxes Mandatory

Sec. 700-9(a). New mandatory rule requires *marking* of all boxes and enclosures containing emergency circuits.

Because circuiting for emergency lighting, exit signs, and emergency power is so critically related to personnel safety, it is important that junction boxes, auxiliary gutters, and other equipment enclosures used for emergency circuits be clearly and effectively "marked" to identify them. Although the new rule does not specify what form the marking should take, it seems clear that the rule can be satisfied by using adhesive labels with, say, black lettering on a bright red background stating something like "CONTAINS EMERGENCY CIRCUITS." (See Figure 1.) The reason for identifying all such emergency circuit enclosures is to alert electricians or others who might work on the circuits to the high priority of keeping the circuits intact and op-

Figure 1 Identification label on this junction box alerts electricians and maintenance personnel to the critical nature of wiring contained in the box—as required by the new Code rule.

erative. So often when building interiors are being altered or redesigned, electricians will be opening boxes, gutters, and other enclosures to revise electrical circuiting. When such work is done over a period of days or weeks, electricians may not restore open circuits to operating condition at the end of the work day if they are going to have to reopen the wiring the next morning. If emergency circuits are left open and an electric power outage occurs during the night in an occupied building, personnel in that building will have their safety jeopardized by failure of emergency light and power and exit signs that are essential to safe, orderly evacuation of the building. Marking on any and all enclosures that contain emergency circuit wiring will caution electricians to "button up" those circuits at the end of each work day, even though the circuits may be reopened the next day. High reliability and continuity of all emergency wiring will be maximized by such warnings. It should be carefully noted that the wording of the rule calls for marking on each and *every* box or other enclosure throughout the emergency lighting and power system of a building. Like a chain depends on each link, integrity of a warning system depends upon marking on every point at which the emergency circuiting might be opened. And that can amount to very many points in the emergency systems used in modern commercial, institutional, and industrial buildings. Manufacturers of adhesive labels make available bright labels with bold lettering to give the necessary warning for this application—e.g., "CAUTION—EMERGENCY CIRCUITS" or "EMERGENCY CIRCUITS WITHIN." Such labels must be placed on enclosures for panelboards, transfer switches, and associated control enclosures, as well as on all boxes, conduit bodies, etc.

Field Marking and Identification of Electrical Installations

The National Electrical Code (NEC) in Sec. 90-1 informs us that the purpose of the Code is the practical safeguarding of persons and property from the hazards associated with the use of electricity. And, that "compliance therewith and proper maintenance will result in an installation essentially free from hazard..." Among the many detailed requirements given by the NEC to assure that an installation is essentially free from hazard are those rules that call for "field marking" of equipment.

Throughout the Code, a variety of references are made to "marking" and "identification." Many of these references apply to the equipment manufacturers; however, the others are intended to require the installer to mark equipment so that people servicing or maintaining this equipment at a future date—typically, operations and maintenance personnel—will be alerted to certain characteristics and details of the installation. Such "field markings" are intended to maximize safety for operations and maintenance personnel, as well as prevent equipment damage during service, maintenance, or repair.

While the importance of these rules is commonly acknowledged, the frequency with which these rules are overlooked is staggering. Inasmuch as failure to provide these markings is a certain and clear Code violation, and that such a violation is exactly the type of violation so often sought by "forensic engineers" when an accident occurs, the installing electrician must be more aware than ever of exactly what is required by these rules and ensure that all the "t's" are crossed and all the "i's" are dotted.

It is worth noting that the number of field markings required by the NEC has again been expanded in the 1990 edition of the NEC. As the NEC goes into great detail to describe the minimum requirements that result in a safe installation, attention to detail on the part of the installer must be exercised. That is, it is not only necessary to *know* what must be done, the installer must ensure that it *is* done. This attention to detail not only applies to the correct sizing of services, feeders, overcurrent protection and other electrical equipment, but also to the marking or other identification of equipment by means of field-installed signs or other field-provided identification.

The following discussion will indicate those sections of the NEC where field marking is required, illuminate the reason for which the marking is required, and give examples of the required marking that will meet the letter and intent of these Code sections.

General considerations

In addition to the specific requirements spelled out in the Code sections that call for field markings, there are certain other considerations that should be discussed.

One consideration that is not specifically addressed by any Code rule deals with the method of marking. That is, when providing field markings, would it be permissible to use, say, a magic marker or must the marking be in the form of a manufactured sign? Another has to do with abbreviations. Are abbreviations or "legends" acceptable forms of field marking? Also, in what language(s) must the equipment be marked? The Code is not entirely clear on these matters.

Generally, where the Code calls for field markings, such markings are required to be legible and durable. This wording has been interpreted to mean that any field marking provided by the installing electrician must be able to be *read* by the people who will operate or service the equipment and the marking must be capable of withstanding the environment in which it is used so that the marking will be able to effectively communicate the warning, caution, or other information over the entire life of the equipment.

To comply with the generally accepted meaning of the word "legible," such markings should be typed or neatly printed. Abbreviations and other shortened forms of expressions, although not prohibited, should be avoided or at least minimized so as to ensure effective communication of the condition or hazard associated with the equipment. And, where abbreviations or other shortened forms of communication are used, all individuals involved with the operation and/or maintenance of the equipment should be trained so that they understand the meaning of the abbreviated markings. Additionally, to facilitate the electrical inspection, such information should be made available to the inspecting authority prior to the actual inspection date.

The question of what language must be used requires a bit of common sense. Obviously, if the individuals who are to operate or maintain the electrical system do not understand English, then markings, warnings or cautions printed in English will not effectively communicate the desired message or information. Therefore, to be completely on the safe side of this question, in those instances where the individuals operating or maintaining the equipment speak a language other than English, the Code-required markings, warnings or cautions should be printed in the language(s) understood by operation and maintenance personnel. While this may seem excessive, it has been repeatedly ruled in court that failure to provide instruction in a language understood by those who must read the instruction (i.e., those who operate and/or maintain the equipment) is akin to not providing the marking, warning, or caution at all. Always ensure that any Code-required marking, warning, or caution is printed in language(s) understood by all persons known to be involved with the operation and/or maintenance of the equipment.

The word "durable" is a bit more difficult to come to grips with. For example, many inspectors will not accept a marking, warning, or caution that is printed with a magic marker. However, there are many instances where such marking has been used and proven to be sufficiently resistant to the environment. The same can be said of Dymo-labels. Nonetheless, as we are all aware, the authority having jurisdiction—generally an electrical inspector—is the final judge of acceptability. Although use of manufactured or fabricated signs is accepted by virtually all inspecting authorities, to avoid any problems during the inspection, it would be most prudent to contact the inspecting agency and discuss the acceptability of the method of marking you intend to employ.

The last concern that should be addressed is the actual wording of the Code-required warning or caution signs. For example, in Sec. 710-43, the Code requires all enclosures rated over 600V that contain "switching" or "control parts" to be marked "WARNING—HIGH VOLTAGE." While such a sign would satisfy the requirement of Sec. 710-43, several rulings rendered in courts across the country have found that such a warning is inadequate. The reason given for these rulings is that, although the sign conveys a warning, it gives the reader no instruction as to what action must be taken with respect to the hazard or condition that exists, and therefore, the warning is incomplete. To "complete" the warning required by Sec. 710-43, additional wording must be added to give the reader instructions with respect to the action to be taken. Here, addition of the words "KEEP OUT" would serve to provide the necessary instructions. It is worth noting that in NEC Sec. 110-34(c), which deals with the marking of rooms or enclosures containing exposed energized parts or exposed conductors operating at over 600V, the recommended wording includes the additional phrase "KEEP OUT." In any case where the Code wording required for a warning or caution sign does not include an instruction for action to be taken by the individual reading the sign, always add such an instruction to limit legal exposure. Remember, the Code-recommended wording is the bare minimum and, in those instances where the Code can be construed as being inadequate,

The warning sign at top in this picture is not complete as it gives no instruction with respect to the action that should be taken. In addition to the fact that the rule of Sec. 110-34(c) requires the additional wording "KEEP OUT," it has been repeatedly ruled in court that such instructional wording must be included in all warning and caution signs, even if a particular Code rule does not specify the inclusion of an instructional statement. Here, another sign that includes the required instructional wording was mounted below the original warning sign to bring this installation into compliance with the letter and spirit of Sec. 110-34(c).

The rule of Sec. 110-34(c) for equipment rated over 600V expands the requirements for marking to include enclosures, as well as rooms or buildings that contain exposed live energized parts. For equipment rated 600V or less, Sec. 110-17(c) only requires the entrance to rooms or guarded locations containing exposed energized parts to be provided with such warning signs.

it is up to the individual engineer, contractor, or operations and maintenance person to exercise good judgment and go beyond that which is minimally required.

Requirements for electrical installations—Article 110

Secs. 110-17(c) and 110-34(c). Two sections of Article 110 require warning signs to be installed at the "entrances to rooms or other guarded locations" where such rooms or locations contain exposed, live electrical parts.

Section 110-17(c), which governs installations rated 600V or less, requires a warning sign at the entrance to any "room or other guarded location" where equipment operating at 50V or more has exposed live parts that could be accidentally contacted and is required to be "guarded" as covered by Sec. 110-17(a). Live parts in general should be protected from accidental contact by complete enclosure—that is, the equipment should be "dead front." But in those cases where this is not done, "guarding," as described in Sec. 110-17(a)(1) through (4), must be provided and a sign must be placed at a "conspicuous" location at the entrance to the room or guarded location so as to prohibit entrance by other than qualified persons. Although no specific wording is given in the Code rule, something to the effect of "WARNING—EXPOSED ENERGIZED ELECTRICAL EQUIPMENT (CONDUCTORS)—AUTHORIZED PERSONNEL ONLY—ALL OTHERS KEEP OUT" should serve to meet the requirement of this section.

Sec. 110-34(c) expands this requirement for systems operating over 600V to include enclosures. This section requires that all buildings, rooms, or enclosures that contain energized parts operating at over 600V be provided with a sign on the door to the building, room, or the enclosure door that reads "WARNING—HIGH VOLTAGE—KEEP OUT." Any additional instructions may be included as the individual sees fit, but the sign must at least contain this Code-suggested wording. It is worth noting that this requirement applies to all such locations covered by the basic rule of Sec. 110-34(c) **and** those covered by the exception to that basic rule.

Sec. 110-22. In Sec. 110-22, the code has long required the identification of disconnecting means for motors, appliances, services, feeders and branch circuits at their point of origin to indicate the purpose of the disconnecting means. This rule goes on to say that such marking is not required where the disconnect is "located and arranged so that the purpose is evident." However, such permission should be ignored

Here's a classic example of a violation of Sec. 110-22. As covered in this section of the Code, ALL disconnecting means for "motors and appliances, and each service, feeder, or branch circuit" must be marked to indicate the load(s) they feed. Even though this installation was inspected and approved (arrow), the fact that the inspector accepted the installation does not release the installing contractor from liability should an accident occur. Always make certain that ALL Code-required markings are provided. Remember, the job's not complete until the required markings are provided.

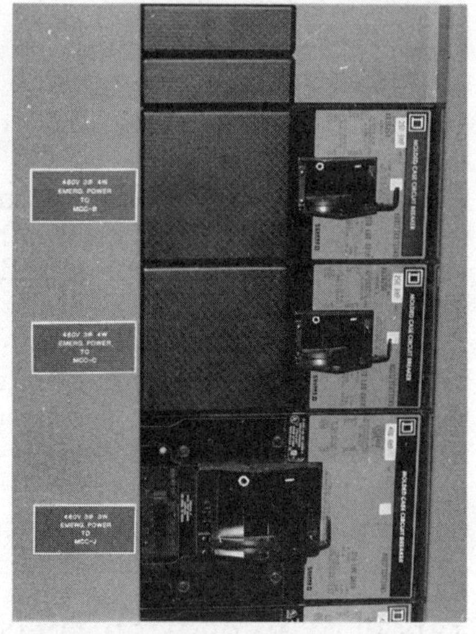

Here's a classic example of an installation that meets the requirements spelled out in Sec. 110-22. The embossed plastic signs identify the loads supplied by each of these feeder breakers.

because although the purpose may be evident to the installing electrician, this rule also serves to provide information to persons occupying the building or facility. Inasmuch as the purpose of a given disconnect may *not* be "evident" to such persons, omitting this marking could expose the installer to legal action in the event of an accident.

As far as the Occupational Safety and Health Administration (OSHA) is concerned, Sec. 110-22 applies to all electrical systems whether new, existing, modernized or altered. As indicated, the basic purpose of this requirement is to assist individuals within the building or facility to identify the source of electrical energy and permit them to readily deenergize a circuit quickly to remove a potential hazard to persons or property. This requirement is all too often overlooked, and the result can be devastating. Such marking should clearly indicate the equipment or circuits fed. In cases where more than one circuit fed from a single panel or enclosure serves similar

Although the NEC gives many requirements for providing markings, warnings, and cautions at specific equipment and locations, no mention is made of what method may be used. Here, in accordance with Sec. 110-22, the disconnects for the individual motor branch-circuits are marked to indicate the load fed by each. As can be seen, five different methods—from pencil (1) to magic marker (5) to Dymo labels (3) to painted lettering (4) and manufactured signs (2)—were employed. Although experience has shown that many inspectors will not accept the use of pencil, magic markers, or Dymo labels, use of painted lettering or manufactured signs has been found to be acceptable to virtually all inspecting authorities. The Code typically only requires that such markings be legible and durable, and that wording is very much subject to interpretation. To avoid any problems during inspection, check with the local inspecting authority to assure acceptability of the method to be used prior to marking the equipment.

equipment, such as lighting, the marking should identify which lighting fixtures are fed by each branch-circuit. All breakers in service equipment, feeder panels, and panelboards, as well as individual disconnect switches and CBs must be fully and clearly identified by field markings on the circuit directories or elsewhere on the equipment. And all such markings of equipment must be substantial and capable of withstanding the environment in which the equipment is installed.

The second paragraph of this section requires the marking of the enclosure containing the "series rated" fuses or CB's to indicate that the fuses or breakers were applied with a series combination rating. That applies when fuses or CBs are applied at a point in the electrical system where the available short-circuit current is in excess of the interrupting capacity of the individual feeder and/or branch circuit fuse or CB, but the feeder and/or branch circuit overcurrent protective device has been tested in combination with the upstream (service and/or feeder) overcurrent protective device to safely interrupt a value of short-circuit current in excess of that for which the individual feeder and/or branch circuit is listed. This is intended to ensure that operation and maintenance personnel will be aware of the fact that the overcurrent protection is part of a "listed system" and that proper operation can only be assured by replacing or adding fuses or breakers that have also been listed as part of the same system. Although the code only requires this warning sign to read: "Caution—Series Rated System" it would be advisable to add an instruction, such as "Identical Component Replacement Required" or similar wording to complete the warning sign and ensure that a non-series-rated component will not be installed either as a replacement or an addition in the future.

Use and identification of grounded conductors—Article 200

Sec. 200-6(a). The basic rule of Sec. 200-6(a) describes the Code-recognized "means" for identifying a grounded circuit conductor (this is generally a neutral conductor, but it may also be a phase conductor, e.g., in corner-grounded delta systems). For conductor sizes No. 6 and smaller, this rule requires a continuous white or natural gray outer covering for the entire length of the conductor. As covered by Exception Nos. 1 and 2, multiconductor varnished-cloth-insulated cables and fixture wires, as covered in Article 402, are exempt from the basic rule. Also, where conditions of maintenance and supervision exist to assure only qualified persons will do work on the installation, it is permissible for grounded conductors in multiconductor cables to be identified at the time of installation by a distinctive white marking or other effective means (Exception No. 3). Exception No. 4 to this section allows for identifying the grounded conductor of Type MI, metal-sheathed cable by distinctive white marking at the terminations during installation.

In 200-6(b), which covers grounded conductors larger than No. 6, the Code permits either the same "means" described in part (a) or a "distinctive white marking" applied during installation at its points of termination. Additionally, multiconductor flat cables No. 4 or larger can employ an external ridge as a means of identifying the neutral.

The Exception to Sec. 200-6(b) permits identification of the grounded conductor in multiconductor cables by a distinctive white marking during installation, but this may be used only where conditions of maintenance and supervision assure that only qualified persons will service the installation.

As covered in Sec. 200-6(d), where conductors of different voltage systems (e.g., 480Y/277VAC and 208Y/120VAC) occupy the same raceway enclosure, auxiliary gutter, etc., provisions must be made to distinguish the neutral conductors from each other. Secs. 200-6(a) and (b) allow the neutral in either system to use a white or natural gray covering. And Sec. 200-6(d) requires the grounded (neutral) conductor of the other system to have a white outer covering with an identifiable colored (other than green) stripe. This rule also applies to feeder and branch-circuit conductors of the *same* voltage that are fed from "different systems," such as where they are fed from different services or transformers and the conductors are installed in the same raceway, conduit, etc.

Branch circuits—Article 210

Sec. 210-4(d). For branch circuits, a new rule has been added to the 1990 NEC for the identification of ungrounded conductors in Sec. 210-4(d). Where two or more nominal voltage systems exist in the same building, not only must a differentiation be made between the grounded (neutral) conductors when the conductors of the different systems run in the same raceway (NEC Sec. 200-6), but also the ungrounded (hot) phase conductors must be identified. The Fine Print Note (FPN) to this section advises that this can be accomplished by color coding, marking tape, tagging or other effective means.

Color coding requirements were removed from the code many editions ago, but are still employed by many electrical design and installation people. Color coding of conductors enhances personnel safety and should be employed virtually all the time. And, as indicated by the FPN, such an approach will serve to satisfy this requirement.

The second part of this rule requires that the means of identification used—whether it be color coding, tagging, marking, etc.—must be "permanently posted at each *branch-circuit* panelboard" to indicate how the ungrounded circuit conductors are identified to comply with this rule. Also, note that identification of ungrounded conductors is required whether or not the conductors of different systems occupy the *same* or *separate* raceways, enclosures, etc. If there is more than one nominal voltage system within a building or facility, then the rules of this section must be satisfied. [Of all the acceptable means available to satisfy this requirement, it is quite clear that color coding of multiwire branch-circuit wires is the easiest and surest method. And such color coding should be voluntarily applied to service, feeder, subfeeder, and other than multiwire branch circuits. Beware that in certain installations use of the widely accepted "BOY"—brown, orange and yellow—system of conductor phase-identification may be prohibited as these colors may be required by other Code rules to be reserved for a particular application. [See Secs. 215-8, 230-56, and 517-160(a).]

Sec. 210-5. Sec. 210-5(a) reiterates the requirements for color coding of the neutral found in Sec. 200-6 (see above) and Sec. 210-5(b) covers color coding of the equipment grounding conductor.

The basic rule of Sec. 210-5(b) requires the equipment grounding conductor used with a branch circuit to be identified by a continuous green color or green with one or more yellow stripes or the equipment grounding conductor may be bare.

Exception No. 1 to this rule permits identifying the equipment grounding conductors at the time of installation as covered in Sec. 250-57(b), Exception Nos. 1 and 3 and Sec. 310-12(b), Exception Nos. 1 and 2. Because these two exceptions to these two sections are virtually identical, the following explanation will focus on the rules of Sec. 250-57(b), Exceptions Nos. 1 and 3.

As permitted by Exception No. 1 to Sec. 250-57(b), insulated or covered conductors in sizes No. 4 and larger may be identified as an equipment grounding conductor, wherever the conductor is accessible by: (1) completely removing the insulation or covering from the entire length of the conductor within the enclosure; (2) by green coloring on the conductor insulation or covering within the enclosure; (3) by marking the conductor insulation or covering with green-colored tape or green-colored adhesive labels within the enclosure.

Exception No. 3 to Sec. 250-57(b) covers the alternative means of identification for multiconductor cables where it is assured that only qualified persons will perform servicing or maintenance on the electrical installation. This exception recognizes the same methods permitted by Exception No. 1 to Sec. 250-57(b).

It is worth emphasizing that whenever equipment grounding conductors are identified in accordance with any of the methods described in the appropriate Exceptions to either Sec. 250-57(b) or Sec. 310-12(b), the means of identification used must be provided at each and every point in the system where the equipment grounding conductor is accessible. Also, the stripping must be provided for the entire length of the conductor that is exposed within the enclosure. For example, if stripping of the conductor is used, the conductor must be stripped back to the point where the equipment grounding conductor leaves the raceway and enters the enclosure. Inasmuch as this must be done at every point in the system where the equipment grounding conductor is accessible—in all service equipment, every switchboard, panelboard, splice box, junction box, device box, conduit body, etc.—it becomes clear that the color coding required by the basic rule is the easiest and least time-consuming method available to provide the required identification.

Feeders—Article 215

Nowhere in Article 215 is there a requirement for color coding or otherwise identifying the ungrounded phase conductors as is required by 210-4(d) for branch circuits.

Sec. 215.8. On a 4-wire delta system where the midpoint of the one phase is grounded—the so-called "high-leg" or "red-leg" delta—to supply lighting, receptacles, and other similar loads, in addition to 3-phase power loads, Sec. 215-8 requires identification of the phase leg with the higher voltage to ground by means of an orange outer finish, tagging, or other effective means. This phase conductor must also be identified at each point in the system where a connection might be made and the system neutral is also available. This includes switchboards, panelboards, motor control centers, junction boxes, and other enclosures to, or through which, the feeder conductors are run. The idea is to prevent someone from accidentally or unknowingly connecting a 120VAC load to the "high-leg," which is 208VAC to neutral.

Although this rule only requires that the high-leg be identified, many installers also mark the various enclosures to indicate what the identification means. That is, a sign or legend is posted within the enclosures that states something to the effect of "The Orange-Colored Conductors are 208VAC to ground" and some even identify the breakers that have a 208VAC potential with respect to the neutral with a piece of orange-colored tape. As indicated, such additional marking is not *required,* but it is certainly good practice to provide such marking as it will serve to maximize the effectiveness of the warning that is intended to be communicated.

A question exists with respect to high-leg delta systems and Sec. 210-4(d). Is such a distribution system considered to be a single nominal voltage system or is it considered to be more than one? Because such systems are typically used to provide power for 3-phase 240VAC loads and 120VAC single-phase loads, should the presence of the 208VAC high-leg be considered another nominal voltage system? The easiest and safest approach would be to assume that such a distribution *does* constitute more than one nominal voltage system and comply with the requirements for identification of *all* ungrounded (phase) conductors as given in Sec. 210-4(d).

If the rules of Sec. 210-4(d) are *not* intended to be applied to the high-leg or red-leg delta systems, it is always good practice to provide conductor identification right down to the branch circuits and their outlets as it could save someone's—usually another electrician's—life. By providing a means of identification throughout the building or facility, even though it is only specifically required for service and feeder conductors, greater safety will be afforded for people, and equipment will be less likely to be damaged by accidental connection of a 120VAC load to the 208VAC high-leg, which is the basic intent of Sec. 215-8. When one considers that the majority of load added to a system after initial construction is typically branch circuit load, common sense would dictate that a method for identification should be carried to the branch-circuit level of distribution. (See discussion under Sec. 230-56.)

Services—Article 230

Sec. 230-2. This section, which covers the number of services permitted for a single building or structure, requires a "directory" to be posted at each service drop or lateral when more than one service is provided in accordance with the exceptions to the basic rule. Although no definition is given for the term "directory" and no wording for this "directory" is suggested by Sec. 230-2, it would seem that this rule would require a sign to be posted that says something to the effect of: "This building (structure) is supplied by two (three, four, etc.) service drops (laterals) in accordance with Exception No. X to NEC Sec. 230-2. The other service drop(s) (laterals) is (are) located on the first floor at the Northeast corner of the building (structure)."

Additionally, a plan view of the building indicating where the reader is standing and where the other service drops or laterals are in relation to that location may be acceptable as the required "directory." A graphic representation of the service locations should also include wording similar to that recommended in the first sentence above for a sign.

Whether a sign or graphic representation of the service locations is used, this marking should be placed on or at the service disconnect in such a location as to ensure that anyone who might be attempting to open the utility feeds to the building or structure will see it.

Sec. 230-56. An identification requirement for service conductors of a "high-leg" delta system, similar to that required for feeder conductors in Sec. 215-8, is given in Sec. 230-56. Again, the phase conductor with the higher voltage to ground—the "high-leg"—must be identified by an orange-colored outer finish or "other effective means," which could permit the use of orange-colored tape, or tags that say something like "high-leg—208V to ground." (See discussion under Sec. 215-8.)

Sec. 230-70(b). Section 230-70(b), which deals with service disconnecting means, requires each service disconnect to be identified as a service disconnect. Wording to the effect of "Service Disconnect" should serve to satisfy this rule. As with all Code required markings, when installed outside, additional consideration must be given to the method used to assure it is capable of withstanding the environment in which it is installed.

Sec. 230-72(a). Sec. 230-72(a) essentially reiterates the basic requirement of Sec. 110-22 for marking of disconnects. This rule states that when two to six service disconnects are used in accordance with Sec. 230-71, they must be grouped and the load they supply must also be

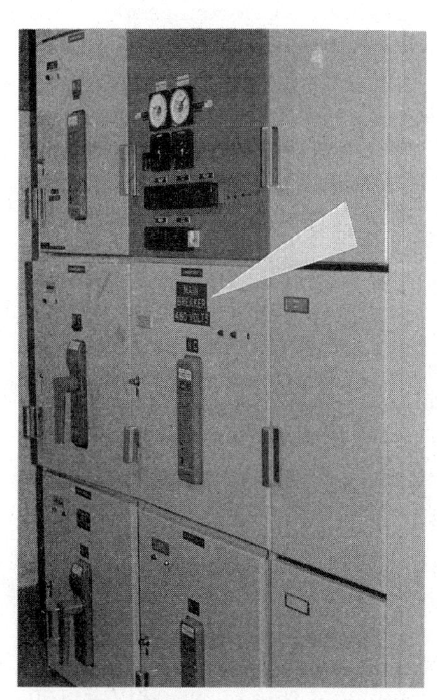

Is this a service disconnect? That's hard to say from the wording used. Although the Code does not specify any particular wording in Sec. 230-70(b), it does require each service disconnect to be "permanently marked to identify it as a service disconnecting means." Use of the words "Service Disconnect" or "Service Main Breaker" would more readily communicate the fact that this breaker is the service disconnecting means.

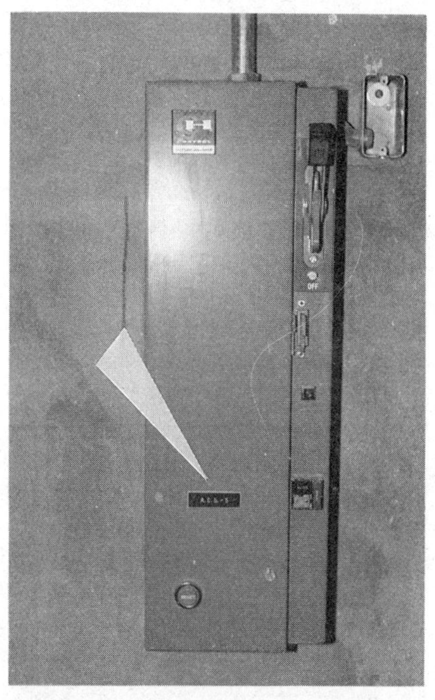

The disconnect shown above is marked with an embossed plastic sign that reads "A.C.S.-5" (arrow). Although not specifically prohibited by the NEC, use of abbreviations or other shortened forms of communication should be avoided or at least minimized to assure that the marking is capable of effectively communicating the desired information. Where abbreviations, acronyms, plan designations, etc., are used to provide the Code-required markings, all personnel that operate and/or maintain electrical equipment should be trained to assure that they understand how to interpret the markings.

indicated on each such disconnect. Because these disconnects must also be marked as service disconnects to comply with the requirements of Sec. 230-70(b), wording to the effect of "Service Disconnect for power and lights in Bob's Candy store" or "Service Disconnect. Feeds Power and Lighting Panel No. 3 in Tenant Space No. 1" or similar wording should serve to satisfy these requirements. It is worth noting that whatever phraseology is used, it should convey as precisely as possible the nature and location of the loads served.

Services exceeding 600 volts, nominal: Secs. 230-200 and 230-203. This section states that all preceding provisions required in this article also apply to service conductors and equipment rated over 600V. This wording has the effect of requiring compliance with all other sections in Article 230 that apply to services rated 600V or less. And this would include the markings required for number of services [Sec. 230-2], identification of service disconnects [Sec. 230-70(b)], and identification of the load supplied where two to six disconnects are used [Sec. 230-72(a)]. Additionally, other markings, as required by Secs. 230-203 and 230-205, must be provided.

Sec. 230-203 essentially repeats the rule of Sec. 110-34(c). Again, signs must be posted to warn unauthorized personnel of the dangerous voltage present within a room, enclosure, or building containing service equipment rated over 600V "where unauthorized persons might come in contact with energized parts." Because the first part of Sec. 110-34(c) generally requires rooms, enclosures, and buildings containing exposed conductors and/or live, exposed parts operating at over 600V to be locked, this last part of Sec. 230-203 could be interpreted as eliminating the need for a warning sign because "unauthorized personnel" can not normally come in contact with energized parts when the room, enclosure, or building is locked. This qualification is *not* presented by the rule of Sec. 110-34(c). That is, Sec. 110-34(c) requires warning signs whether the enclosure, room, or building is locked or not (as permitted by the Exception). And the warning sign is required without any qualification with regard to unauthorized personnel and the possibility (the Code rule uses the word "might") of their contacting energized parts. Is the rule of Sec. 230-203 a modification of the general requirement given in Sec. 110-34(c)? Would locking of the enclosure, room, or building containing service equipment rated over 600V be considered as preventing unauthorized personnel from coming in contact with energized parts and thereby eliminate the need for a sign?

To be on the safe side, it is best to always assume that "unauthorized personnel might come in contact with energized parts" and provide the warning sign called for in Sec. 230-203 even if the room, enclosure, or building is locked.

Sec. 230-203 suggests the use of the wording: "Danger—High Voltage—Keep Out." This is the Code-prescribed minimum. Additional wording, as may be deemed necessary by the engineer or installer, may be added. And this sign must be "posted in plain view." Posting of this warning at eye level on the door(s) for room, building, or enclosure should serve to satisfy this requirement.

Sec. 230-205(a), Exception. Generally, the service disconnects for services rated over 600V must be located as required by Sec. 230-70 or Sec. 230-208(b) and they are required to comply with the marking requirements given elsewhere in Article 230. The Exception to the basic rule of Sec. 230-205(a) applies where multiple buildings are "under single management." It permits the service disconnecting means for one building to be located remotely in another building provided a control device is installed in the building to be supplied to open the disconnect. This control device must be located "as near as practicable" to the point where the "service conductors" enter the outbuilding. Although the term "service conductors" is incorrectly used (such conductors would actually be "feeder conductors" according to Code definition), it is clear enough that this device is required to be installed near the conductor entrance point. And the control device must be marked "to identify its function." A marking on or immediately adjacent to—above, beside, or below—the control device must be provided and it should say something to the effect of "Local Service Disconnect Control Device" or "Local Operator for Remote Service Disconnect" or other wording that conveys the idea that the device operates a remotely located service disconnect. Additionally, the control device is required to provide a status indication of the service disconnect. Although this device itself is only required to "open" the service disconnect, the status indicator must provide indication of both the "on" and "off" positions.

Grounding—Article 250

Sec. 250-57(b). The last sentence in the basic rule of Sec. 250-57(b) requires a covered or insulated equipment grounding conductor to be identified by a continuous green color or green with one or more yellow stripes or the equipment grounding conductor must be bare. This is essentially the same requirement presented by the basic rule of Sec. 210-5(b), but this rule would cover *all* equipment grounding conductors not just those run with branch circuits.

As permitted by Exception No. 1 to Sec. 250-57(b), insulated or covered conductors in sizes No. 4 and larger may be identified as an equipment grounding conductor, wherever the conductor is accessible by: (1) completely removing the insulation or covering from the entire length of the conductor within the enclosure; (2) by green coloring on the conductor insulation or covering; (3) by marking the conductor insulation or covering with green-colored tape or green-colored adhesive labels.

Exception No. 3 to Sec. 250-57(b) covers the alternative means of identification for multiconductor cables where it is assured that only qualified persons will perform servicing or maintenance on the electrical installation. This exception recognizes the same methods permitted by Exception No. 1 to Sec. 250-57(b) for identifying an equipment grounding conductor that is larger than No. 6.

It is worth emphasizing that whenever equipment grounding conductors are identified in accordance with any of the methods described in the Exceptions to Sec. 250-57(b), the means of identification used must be provided at each and every point in the system where the equipment grounding conductor is accessible. Also, the stripping must be provided for the entire length of the conductor that is exposed within the enclosure. For example, if stripping of the conductor is used to identify the equipment grounding conductor, the conductor must be stripped back to the point where the equipment grounding conductor leaves the raceway and enters the enclosure. (Fig. 1) Inasmuch as this

EXAMPLE

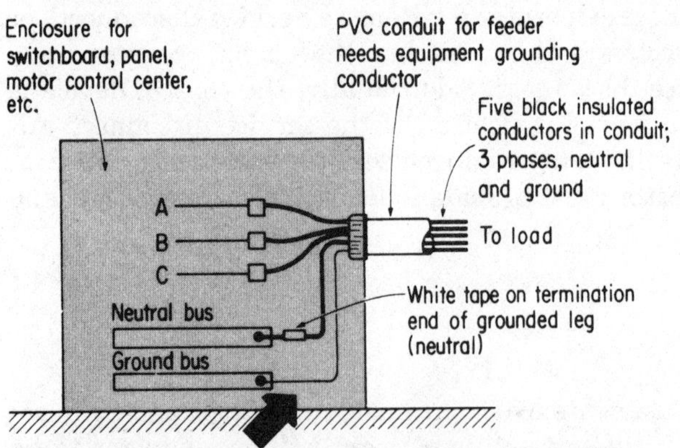

Black insulated conductor used as equipment grounding conductor has all insulation stripped from entire length exposed in enclosure

Figure 1 A conductor, larger than No. 6, with any colored covering may be used as an equipment grounding conductor provided the covering is completely stripped back to the point where the conductor enters the enclosure (Sec. 250-57).

must be done at every point in the system where the equipment grounding conductor is accessible, it becomes clear that the color coding required by the basic rule is the easiest and least time-consuming method available to provide the required identification.

Conductors for general wiring—Article 310

Sec. 310-12. The first sentence of Sec. 310-12(a) presents the same rule as that given in the first sentence of Sec. 200-6(a). And the second sentence of Sec. 310-12(a) repeats the rule given by the second sentence of Sec. 200-6(b). And, in addition to recognizing exceptions for the same applications covered by the exceptions to Sec. 200-6(a), Exception No. 4 to Sec. 310-12(a) permits identification of the grounded conductor as described in Sec. 210-5(a).

In English, Sec. 310-12(a) says that grounded circuit conductors No. 6 and smaller must be identified by an outer covering or insulation that is white or natural gray. For multiconductor cables No. 4 or larger, the means of identification can be provided by use of a raised ridge on the exterior of the grounded conductor. Multiconductor varnished-cloth cables; fixtures wires, as covered in Article 402; and mineral-insulated, metal-sheathed cable of any size are exempt from this requirement. And, where it is assured that only qualified personnel will service and maintain the installation, the grounded conductor in multiconductor cables of any size may be identified at their termination points at the time of installation by means of a distinctive white marking "or other equally effective means." (This last statement is not very clear and is subject to interpretation by the local inspecting authority.) Additionally, the grounded circuit conductor—usually a neutral, although it can also be a phase leg as in a corner-grounded delta system—may be identified as covered in part (a) of Sec. 210-5. [See discussion under Secs. 200-6(a), (b) and (d) and Sec. 210-5(a).]

Sec. 310-12(b) covers the acceptable methods for identifying equipment grounding conductors. These rules are virtually identical to those put forth by Sec. 250-57(b). [See discussion under Sec. 250-57(b).]

Sec. 310-12(c) requires ungrounded (phase) conductors to be "clearly distinguishable" from the grounded conductor and equipment grounding conductors. This essentially means that phase conductors can be any color other than white, green, white with a colored stripe, or green with a colored stripe.

Underfloor raceways; cellular metal floor raceways; and cellular concrete floor raceways—Articles 354, 356, 358

Secs. 354-9, 356-8, and 358-8. All of these raceways are required to have "markers" installed to facilitate locating the raceways, or a portion thereof, after the installation has been completed. For underfloor raceways, Sec. 354-9 requires a "marker" at the end of a run to identify the last point in the raceway system where an insert may be installed. As covered in Secs. 356-8 and 358-8, for cellular metal underfloor raceways and cellular concrete floor raceways, respectively, "a suitable number of markers" must be provided along the entire length of the run to indicate where the cells are located.

The "markers" used for these raceway systems are usually special flathead brass screws that are screwed into the top of the cells, flush with the floor. As stated, the intent is to assist in locating the cells themselves or, with underfloor raceways (Sec. 354-9), to assist in locating the last point at which an insert may be installed.

Boxes and fittings—Article 370

Under Part D of NEC Article 370, which covers "Pull and Junction Boxes Used on Systems Over 600 Volts, Nominal," the Code again requires the "DANGER—HIGH VOLTAGE—KEEP OUT" marking on medium- and high-voltage (over 600V) pull and junction boxes. However, the rule of Sec. 370-52(e) presents more detail than is generally found in Code rules relating to field markings. Not only does this rule present wording that is very specific in nature, Sec. 370-52(e) also specifically requires the warning sign to be marked on the outside cover and in "block type" that is at least ½-in. high.

To comply, all of these requirements must be satisfied. A sign with the wording "DANGER—HIGH VOLTAGE—KEEP OUT" in ½-in.-high (minimum) block-type lettering must be posted on each and every pull box, junction box, splice box, conduit body, etc., in high- and medium-voltage systems. (Fig. 2) And this marking must be "readily visible." Because the basic rule requires that the cover on such boxes must be marked and that the marking must be readily visible, if ever the marking on the cover *is not* readily visible, mark the cover and also that portion (side, top, bottom) of the box where the warning will be readily visible. This will assure compliance with the "letter" and "spirit" of the rule.

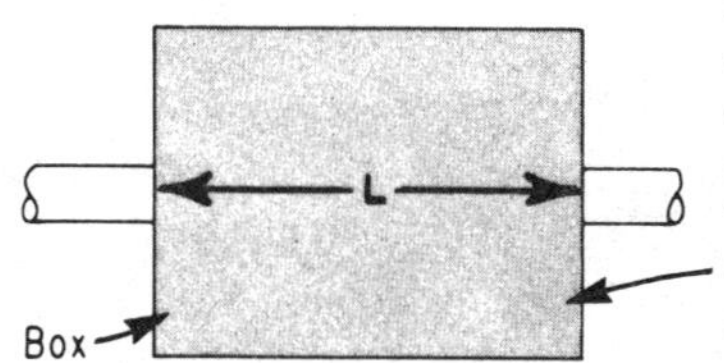

Figure 2 For each and every box containing conductors operating at over 600V, nominal, Sec. 370-52(e) requires that the box *cover* be marked on the outside, "Danger—High Voltage—Keep Out."

Switches—Article 380

In Sec. 380-6(c), the Code puts forth a general requirement that the load-side terminals on single-throw knife switches must be deenergized when the switch is in the "open" position. The Exception to this rule recognizes that electrical systems with UPSs, transformer secondary ties, or emergency or backup generators, may produce an electrical backfeed and cause the load-side terminals to be energized even when the switch is "open." For applications where an electrical backfeed may result in energized load-side terminals when the switch is in the "open" position, the Exception requires a warning sign that reads: "WARNING—LOAD SIDE OF SWITCH MAY BE ENERGIZED BY BACKFEED." (Fig. 3) Additional wording to the effect of "Verify Load-Side Terminals Are Deenergized Prior To Working On This Switch" should be added to provide instruction as to what action must be taken.

Sec. 380-6(c) and its Exception apply to single-throw "knife switches." And the Underwriters Laboratories indicate that switches listed as "knife switches" are the "open-type knife switch." What about safety switches or breakers used at a point in the system where a backfeed may energize the load-side terminals when the safety switch or CB is in the "open" position? What about CBs that are intentionally backfed, such as when feeding a panelboard? Do these applications require a warning sign?

Although no such requirements exist, a new section in the 1990 NEC seems to address the same concern for main CBs and main lug assem-

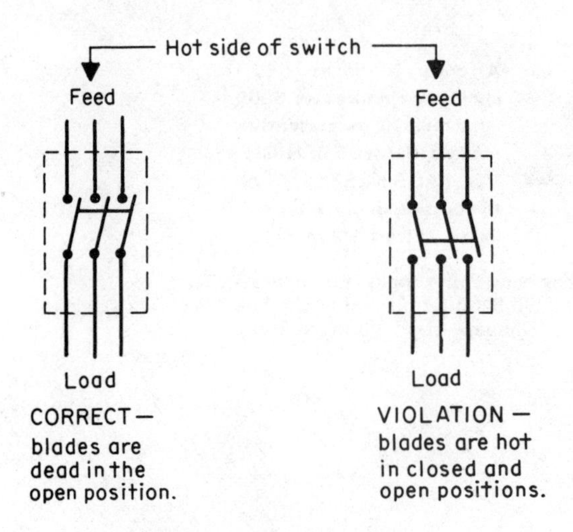

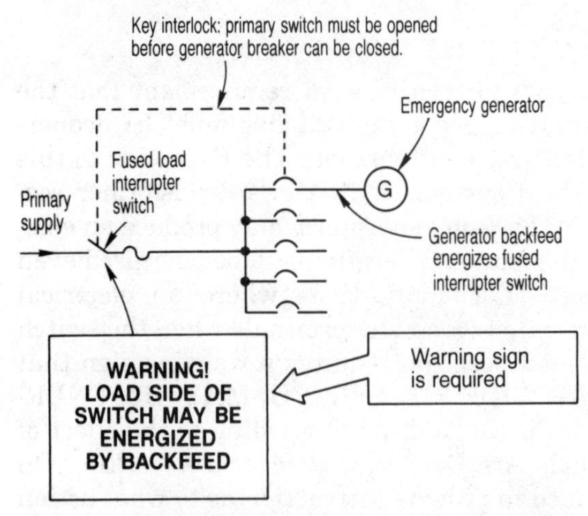

Figure 3 Backfeed of "knife switches" is permitted where the equipment is carefully marked as shown. Although this rule literally applies only to "knife switches" (and UL indicates that switches listed as "knife switches" are the "open-type knife switch with or without fuseholders"), this rule should be extrapolated and such marking should be provided at "other" switches to alert operations and maintenance personnel of the potential backfeed hazard (Sec. 380-6).

blies where they are intentionally backfed to supply a panelboard. Sec. 384-16(f) now requires "an additional fastener" for plug-in type breakers and lug assemblies in panelboards where the CBs or lug assemblies backfeed the panel busbars. The intent behind requiring an additional means of fastening to secure plug-in type CBs and lug assemblies in such applications is that such additional fastening will tend to cause operations and maintenance personnel to take a second look at what they're doing when they attempt to unplug the CB or lug assembly and it doesn't come right out. And during that second look, it is believed that they will realize the CB or lug assembly is not a branch-circuit device, but rather the panelboard main. Although the intent behind this requirement is proper and well intentioned, this rule does not seem to go far enough, in that posting of a warning sign at such locations would better serve the intent of this rule.

Even though the rule of Sec. 380-6(c), Exception, only literally applies to open-type knife switches, it would seem prudent to extrapolate the application of this rule to include safety switches, molded-case switches, circuit breakers, etc., where such devices may be intentionally or unintentionally energized by an electrical backfeed. Voluntary application of this rule will definitely serve to provide a safer working environment for electrical operations and maintenance personnel.

Switchboards and panelboards—Article 384

Secs. 384-3(e) and (f). Sec. 384-3(e) puts forth a requirement for switchboards and panelboards that is similar to other rules that call for marking of the "high-leg"—the phase conductor that is 208VAC to ground—in "high-leg" or "red-leg" delta systems. The rule of Sec. 384-3(e) recognizes marking of either the busbar or the conductor that is at 208VAC-to-ground. And Secs. 215-8 and 230-56, which cover marking of feeder and service conductors used with high-leg systems, require the high-leg *conductor* to be permanently identified. With this in mind, it becomes clear that marking of the conductors is the best method to achieve compliance with Sec. 384-3(e), as well as Secs. 230-56 or 215-8. The marking or means of identification is required to be either an orange outer finish, or orange taping, or "other effective means." [See discussions under Secs. 215-8 and 230-56.] (Fig. 4)

Another consideration for high-leg systems is covered by the second sentence of Sec. 384-3(f). Here the Code requires the high-leg to be terminated on the "B" or center busbar in every panelboard and switchboard fed from such a system. The only exception to this requirement applies to service equipment in the same enclosure as the utility meter. This permission recognizes that many utilities require the "C" or right-hand busbar to be reserved for use by the high-leg. In order to prevent conflict between the basic rule and the utility requirement, this exception permits the busbar arrangement to be other than that described in the second sentence of Sec. 384-3(f). But only within a single- or multi-section switchboard or panelboard enclosure that also houses the utility meter. And the phase arrangement for busbars in

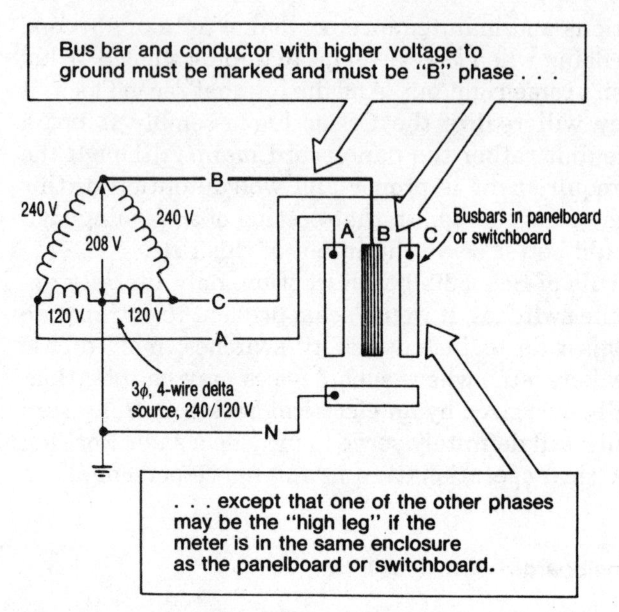

Figure 4 Identification of the "high leg" on a 4-wire delta system is required at all switchboards and panelboards. As noted in Sec. 384-3(e), this may be accomplished by marking the busbar *or* the conductor that has the higher voltage to ground.

all other switchboards and panelboards must comply with the basic rule that calls for use of the center busbar for the high-leg. Therefore, when conductors leave the switchboard or panelboard where the utility metering equipment is installed, the "high-leg" must be terminated on the "B" (center or middle) busbar in every other switchboard and/or panelboard within the system.

The rule of Sec. 384-3(f) also states that busbars in all 3-phase switchboards and panelboards must be connected to a specific phase. The phases are to be arranged in sequence (i.e., A, B, C) and connected from left-to-right, top-to-bottom, or front-to-back, depending on the arrangement of the switchboard or panelboard busbars. (Fig. 5) This is normally done by arbitrarily designating one phase as the "A" Phase at the service equipment. Then, using a phase rotation meter, the "B" and "C" Phases are identified and terminated at the appropriate busbars within the service equipment. This is the easy part. The problem begins when conductors are pulled to the various switchboards and panelboards located elsewhere within the system.

As indicated, this rule requires that each and every right-hand, top, or front busbar in every switchboard and panelboard within the electrical system be connected to the same "A" Phase designated at the

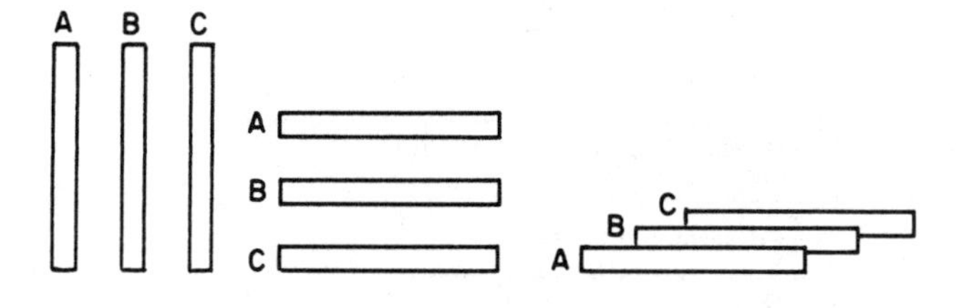

Figure 5 Phase sequence of busbars in every switchboard and panelboard must be fixed (Sec. 384-3).

service equipment. If color coding or some other method of identifying the conductors is not used, then determining which conductor is the "A" Phase becomes more complicated and time-consuming.

The use of color coding on such conductors, even though not specifically required, will serve to assure compliance with the rule of Sec. 384-3(f) and eliminate the need to "ring out" the conductors after they have been pulled. On larger-size conductors, if colored tape (or tagging, or other method) is used to identify the phase from which each conductor is fed, consideration should be given to using more than one piece of colored tape (or tag, etc.) on each conductor. By doing so, if one piece of tape should fall off when the conductors are pulled, there will be at least one other piece of colored tape to identify the phase from which the conductor is fed.

Sec. 384-13. The requirement of Sec. 384-13 is one of the most often overlooked requirements in the Code. Here the Code clearly indicates that failure to completely and legibly mark the circuit directory to identify the loads fed from each circuit within every panelboard is a violation of the NEC. (Fig. 6) This includes new work as well as modifications on existing systems. It is worth noting that OSHA makes this requirement "retroactive." That is, in existing systems, if the panelboard directory is *not* completely and legibly filled out, then it is up to the operations and maintenance personnel to do so. [See discussion under Sec. 110-22.]

As indicated, the basic rule in Sec. 384-13 requires the circuits to be "legibly identified." Too many times the detail of "legibility" is neglected. In these times when liability suits are being initiated at a staggering rate, failure to comply with this simple requirement is extremely imprudent. If your handwriting looks like "chicken scratch," don't fill out a panelboard directory by hand. Have it typed. Each of us must make our own decision with respect to the requirement for leg-

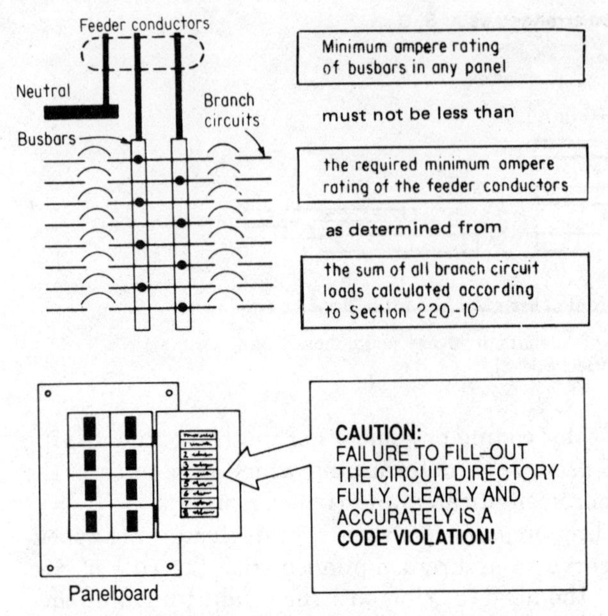

Figure 6 Identification of the loads and functions of all circuits *and* any modifications to the circuits must be provided in every panelboard's circuit directory. Sec. 384-13 further requires the directory to be "located on the face or inside of the panel door."

ibility. Think it over now, because the courtroom is no place to find out that your handwriting is not "legible."

Fixed electric space heating equipment—Article 424

Sec. 424-19. This section covers the NEC requirements on the disconnecting means for fixed electric heating. Basically, a "means" must be provided to disconnect the heater, motor controller(s) and the supplementary overcurrent protective device(s). Where more than one "means" for disconnect is provided, this rule also requires the disconnecting "means" to be "grouped and identified." For example, if one, 3-phase, 480V circuit supplies the resistance heating elements and another 3-phase, 480V circuit supplies the blower motor and controller, or if a separate 120V circuit provides control power to the motor controller, then the disconnects for all circuits must be "grouped"—in the same enclosure or separate enclosures installed adjacent to each other—and marked "Heater Element Disconnect" and "Blower Motor and Controller Disconnect" or "Control Circuit Disconnect," or something that will indicate the purpose of each disconnect grouped at that location. Additional wording might be provided to indicate that all of the disconnects must be opened to completely isolate "the heating elements, the motor controller(s), and the supplementary overcurrent protection."

Fixed outdoor electric de-icing and snow-melting equipment—Article 426

Sec. 426-13. Where outdoor electric de-icing or snow-melting equipment is used, a caution sign must be posted "where clearly visible" to make known that such equipment is present and, thereby, assure safety and prevent interruption of its service. Sec. 426-13 is intended to alert anyone who comes into an area where such equipment is used of the presence and purpose of the equipment. A caution sign saying something to the effect of "Caution—Electric De-Icing Equipment In Use—Do Not Dig in This Area Without Contacting the Building Manager" or similar wording should serve to satisfy this requirement. The location and number of sign(s) necessary is somewhat more difficult to ascertain. For larger areas, it would seem that more than one caution should be provided. Due to the lack of specifics within this Code rule, the best policy is to check with the local inspecting authority for guidance.

Fixed heating equipment for pipelines and vessels—Article 427

Sec. 427-13. In Sec. 427-13, a marking requirement similar to that given in Sec. 426-13 is presented for electrically heated pipelines. Again, in the interest of safety and to prevent disruption of service, a caution must be posted indicating the presence of "electrically heated pipelines or vessels or both." A sign or marking saying something to the effect of "Caution—Burn Hazard—Electrically Heated Pipeline (Vessel or both)—Do Not Touch" would seem to adequately address the concern here. Such signs or markings are to be provided "at frequent intervals" along the pipeline or on the vessel. As was the case with Sec. 426-13, this section (Sec. 427-13) does not clearly indicate the exact number of, or location for, the required caution signs. For that reason, the best policy is to consult the local inspecting authority to assure acceptance of the completed installation.

Motors, motor circuits, and controllers—Article 430

Sec. 430-74(a), Exception No. 1. The second sentence of the basic rule in Sec. 430-74(a) permits the use of a separate disconnecting means to deenergize the motor control circuits provided the disconnect used to isolate the motor and motor controller is located "immediately adjacent" to the separate motor control circuit disconnect. Exception No. 1 to the basic rule of Sec. 430-74(a) is aimed at industrial-type motor control hookups that involve extensive interlocking of control circuits for multimotor process operations or machine sequences.

In recognition of the unusual and complex control conditions that exist in many industrial applications—particularly process industries and manufacturing facilities—Exception No. 1 to Sec. 430-74(a) alters the basic rule that disconnecting means for the control circuits and the motor/motor controller must be located "immediately adjacent one to the other." When a piece of motor control equipment has more than 12 motor control conductors associated with it, remote locating of the disconnect means is permitted under certain conditions. This permission is limited to applications where only qualified persons have access to the live parts and a warning sign is "permanently located" on the outside of the enclosure that indicates the location of the disconnects and identifies the disconnects. This requirement may be satisfied by use of a sign that reads something to the effect of "WARNING—DISCONNECTS FOR CONTROL CIRCUIT POWER ARE REMOTELY LOCATED IN PANEL NO. XX.—DEENERGIZE CIRCUITS 4, 6, 8, & 10 BEFORE WORKING ON THIS EQUIPMENT." Although no specific wording is given, the rule requires that the sign indicate:

1. That the control circuit disconnects are remotely located.
2. The location of the remote disconnects.
3. Specific identification of each remote disconnect.

Make certain that whatever wording is used, these three requirements are covered. (Fig. 7) Inclusion of a requirement to deenergize such circuits prior to working on the equipment will serve to provide an instruction for action to be taken. As always, be sure the method of marking can withstand the environment to which it is exposed.

Sec. 430-102(a). The basic rule of Sec. 430-102(a) says that the disconnecting means for a motor must be "in sight" of the controller location. Exception No. 1 to this rule applies to circuits rated over 600 volts and permits a lock-open type switch or CB disconnect that is not "in sight"

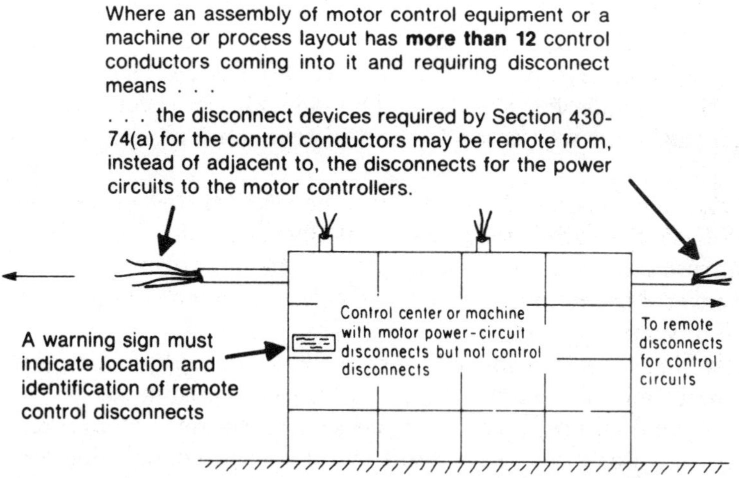

Where an assembly of motor control equipment or a machine or process layout has **more than 12** control conductors coming into it and requiring disconnect means . . .

. . . the disconnect devices required by Section 430-74(a) for the control conductors may be remote from, instead of adjacent to, the disconnects for the power circuits to the motor controllers.

A warning sign must indicate location and identification of remote control disconnects

Control center or machine with motor power-circuit disconnects but not control disconnects

To remote disconnects for control circuits

Figure 7 For extensive interlocked control circuits, disconnects for control power do not have to be adjacent to power disconnects, as is generally required by the basic rule of Sec. 430-74(a). But, a warning sign must be provided. (Sec. 430-74, Exception No. 1).

(50 ft or closer and visible) from the controller location if the controller is marked to identify the disconnect and its location. That is, the warning sign must tell where the disconnect is and how to identify it. (Fig. 8)

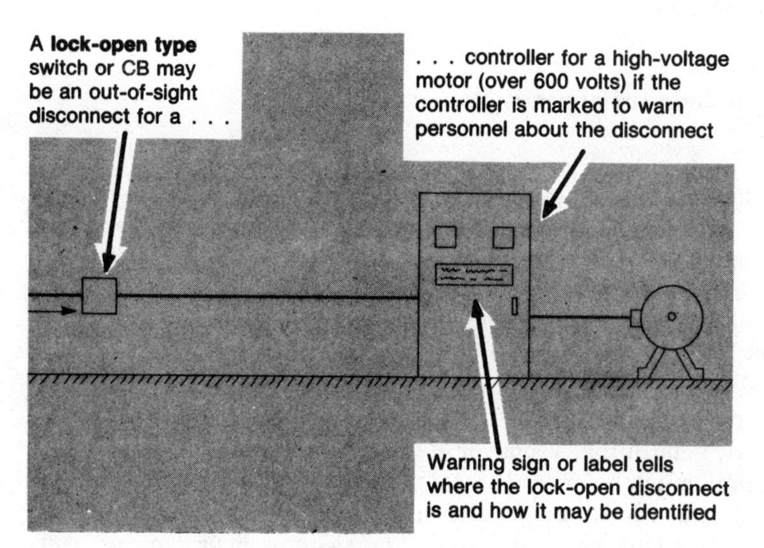

Figure 8 For a high-voltage (over 600V, nominal) motor, a disconnect that is "out of sight" from the controller location may serve as the required disconnect. (Sec. 430-102, Exception No. 1).

Use of wording to the effect of "WARNING—DISCONNECT MEANS IS LOCATED IN EQUIPMENT ROOM NO. XX. OPEN AND LOCK DISCONNECT NO. 9 PRIOR TO WORKING ON THIS EQUIPMENT" should be acceptable and adequately cover the required information as well as provide an instructional phrase.

Sec. 430-109. The type of disconnecting means permitted to be used as a motor disconnect are listed in Sec. 430-109. They are: a motor circuit switch rated in hp, a CB, or a non-automatic molded case switch. There are several exceptions to this basic rule but only one requires a marking of any kind.

Exception No. 4 to Sec. 430-109 sets the maximum hp-rating required for motor-circuit switches at 100hp. Higher-rated switches are now available and will provide additional safety. But for motors rated over 100hp, the Code does not require that the disconnect have a horsepower rating. This exception to the basic rule permits use of an ampere-rated switch or isolation switch, provided the switch has an ampere rating at least equal to 115% of the motor full-load current, as given in Tables 430-147 through 430-150. Where this permission is exercised, the switch must be marked to indicate that it is not to be opened under load. This section specifically calls for use of the wording "Do not operate under load." Use of additional wording saying something like "Non-horsepower-rated disconnect" before the required instructional wording will serve to clarify why this disconnect should not be operated under load.

Transformers and transformer vaults—Article 450

Secs. 450-8(c) and (d). These two sections essentially reiterate the requirements given in Secs. 110-17 and 110-34. Guarding of exposed live parts on transformers is covered in Sec. 450-8(c). This Code section refers the reader back to Secs. 110-17 and 110-34, which explain the guarding requirements for systems under and over 600V, nominal, respectively. Both sections generally require equipment to be enclosed or guarded as described in the appropriate section (either 110-17 or 110-34). And where equipment rated 600V or less is guarded by installation in a building, room, or guarded location a "conspicuous" warning sign must be provided at the entrance to such buildings, rooms, or guarded locations to alert personnel of the presence of exposed live parts [see Sec. 110-17(c)]. For systems operating at over 600V, Sec. 110-34 also requires a warning sign to be provided on enclosures, as well as buildings or rooms, containing exposed live parts.

Part (d) of Sec. 450-8 puts forth an additional marking requirement for transformers. A sign indicating the operating voltage of exposed live parts must also be provided. A single sign may be acceptable to cover both requirements. Combining the requirements of parts (c) and (d), it would seem a warning sign posted on the door to the room, building, structure, enclosure, etc., and reading "DANGER—HIGH VOLTAGE—5000VAC—KEEP OUT" should satisfy the requirements for both Sec. 450-8(c) and (d). (Fig. 9)

Signs or other visible markings must be used on equipment or structure to indicate the operating voltage of exposed live parts

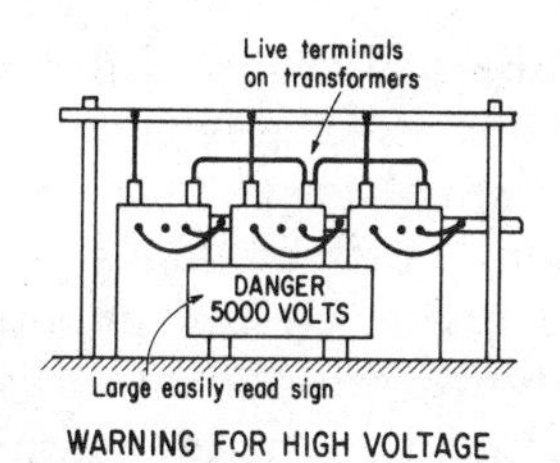

WARNING FOR HIGH VOLTAGE

Figure 9 In addition to requiring guarding of exposed live parts (as described in Secs. 110-17 and 110-34), Sec. 450-8 requires a warning sign that states the operating voltage of the transformer's "exposed live parts."

Sec. 450-28. Askarel or mineral oil transformers that are drained and refilled with another liquid dielectric must be marked as such and must satisfy all rules of its retrofilled status. This rule is intended to maintain safety in all cases where askarel or mineral oil transformers are drained and refilled to eliminate PCB hazards. Marking must show the new condition of the unit. That is, it must identify the type of dielectric now used. For example: "This Transformer has been Retrofilled with a Less Flammable Liquid Insulator" or "This Transformer has been Retrofilled with a Silicone Fluid Insulator." Indication of the specific type of replacement dielectric would be more desirable and better meet the intent of this rule.

Special occupancies: intrinsically safe systems—Article 504

Sec. 504-80(a). The first marking required appears in Article 504 covering "Intrinsically Safe Systems," which was added in the 1990 edition of the NEC. These systems are made up of apparatus that is incapable of releasing sufficient energy to ignite a flammable or combustible atmosphere and that is "approved" for the application.

In Sec. 504-80(a), the NEC requires "identification" of intrinsically safe circuits at "terminals and junction locations." While the word "terminal" is readily understood, there is some question as to exactly what a "junction location" is.

Is it a point on a piece of *equipment* where interconnecting conductors are attached? Is it a point in the *system* where interconnecting conductors would be joined—a junction box? Without any definition to provide guidance, the safest approach would be to consider a "junction point" to be any enclosure, junction box, etc., where conductors are spliced or otherwise joined within an intrinsically safe system. And, provide a means of identification as required by this section at every such location.

"Raceways, cable trays and open wiring" used with intrinsically safe systems are also required to be identified by Sec. 504-80(b). Although no specific method is mentioned for identification of terminals and "junction points" in part (a), Sec. 504-80(b) calls for identification "with permanently affixed labels" at intervals not to exceed 25 ft that say "Intrinsically Safe System or equivalent." Because a label must be used to identify raceways, cable trays, and open wiring, use of labels for "terminations and junction points" would serve to standardize the means of identification and help minimize confusion.

Watch out for the information put forth in Fine Print Note (FPN) 1. It states:

> Color coding may be used to identify intrinsically safe conductors if the color is light blue and no other conductors are colored blue.

This appears to permit use of a specific color-coding scheme as the means of identification. However, as given in Sec. 110-1, FPNs are "explanatory material" and are not Code rules. Because no specific method is given in part (a) of Sec. 504-80, it should be acceptable to use the method of identification described in FPN No. 1 to Sec. 504-80 for intrinsically safe circuits at "terminal and junction points." However, be aware that use of this method to identify cable trays and open wiring in intrinsically safe systems would be a violation of the mandatory requirement given in Sec. 504-80(b)—i.e., "shall be identified with permanently affixed labels." It is worth noting that *anytime* a FPN note appears to give permission that is in conflict with a mandatory requirement given in a basic rule, the mandatory requirement—which is designated by use of the word "shall"—must be satisfied.

These markings are intended to facilitate inspection, as well as alert individuals performing maintenance or repair that they are dealing with a special system and, thereby, prevent accidental intermixing of non-intrinsically safe equipment and conductors. Although Sec. 504-80(b) only requires the identifying labels to say "Intrinsically Safe Wiring" or equivalent (whatever that is), in order to assure that the sign is "complete"—that is, gives a warning *and* instructions with respect to what action is to be taken—in addition to the Code-prescribed wording, such labels should include an instruction, such as "Equipment and conductors used in this cable tray (panel, enclosure, etc.) must satisfy the requirements of Article 504 of the National Electrical Code." This will better serve to achieve the desired intent, which is to prevent the accidental intermixing of "other" equipment or wiring.

Aircraft hangars—Article 513

Sec. 513-6(c). In aircraft hangars, there is a marking required by Sec. 513-6(c) for certain "mobile-type" equipment used for maintaining and servicing aircraft. Specifically, mobile stanchions containing electric equipment that are not located within or likely to be located within the "vicinity of aircraft"—an area that is measured horizontally in all directions for a distance of 5 ft and that extends from the floor to a height 5 ft above the wing's upper surface or the upper surface of the engine enclosure as defined in Sec. 513-2(c)—are required to have a warning sign permanently affixed. This sign must read: "WARNING—KEEP 5 FT CLEAR OF AIRCRAFT ENGINES AND FUEL TANKS."

Although newly manufactured equipment of this type will normally have such a sign attached, any such equipment that does not have a warning sign should be field-equipped with one.

Health care facilities—Article 517

Sec. 517-160(a)(5). Isolated power systems are power systems that are intentionally operated ungrounded to provide an additional measure of safety in combustible anesthetizing areas of health care facilities by preventing ignition of the flammable gases by arcing from a ground-fault. In Sec. 517-160(a)(5), the NEC gives a requirement for identifying the branch circuit conductors used with "isolated power systems."

Sec. 517-160(a)(5) requires the branch circuit conductors in a 2-wire isolated power system to be identified as "orange" and "brown." And for a 3-wire system, the third conductor must be identified "yellow."

It is worth noting that, although use of conductors with insulation that is colored orange, brown, or yellow throughout its entire length would certainly be acceptable, it seems that the wording would also permit "identifying" the branch circuit conductors with orange-, brown-, or yellow-colored tape or with a label that says "orange," "brown," or "yellow" at all locations where the conductors are accessible.

Theaters and similar locations—Article 520

Sec. 520-27(c). Sec. 520-27(c) calls for any "single primary stage switchboard (dimmer bank)" to be marked with a "label" that identifies the number and location of the disconnects for each feeder supplying the stage switchboard. This rule recognizes the fact that the amount of load required by these stage switchboards today is typically going to necessitate the use of multiple feeds. And although this rule does not require the disconnect for each feeder to be installed within the stage switchboard, it does call for identification of the number and location(s) of the feeder disconnects where more than one feeder supplies the stage switchboard. For example, a sign permanently posted in a conspicuous location on the front of the stage switchboard might say, "This switchboard is supplied by two feeders. The disconnects for these feeders are in panel LP-12 located in basement B2."

Motion picture projectors—Article 540

Sec. 540-11(b). In general, Sec. 540-11(b) only permits equipment that is directly associated with the operation and control of "projectors, sound reproduction, flood or other special effects lamps, or other equipment" to be installed within the projection room. But, as covered by Exception No. 1 to Sec. 540-11(b), installation of such "other equipment" is permitted within projection rooms that are "approved for use only with cellulose acetate (safety) film." The second sentence of Exception No. 1 goes on to say that a minimum of two (2) signs must be provided for each such projection room. At least one inside the room (in a "conspicuous location") and one on each door to the room. Both signs are to read: "Safety Film Only Permitted in This Room." Such wording would seem to be complete as it stands because it is simply conveying a command.

NFPA Issues Errata for 1990 Edition of the National Electrical Code (NFPA 70)

The **National Fire Protection Association** (NFPA) recently issued corrections for the First and Second Printings of the 1990 National Electrical Code. These corrections, or "errata," were published in the August/September 1990 issue of the "NFPA Standards Action." As stated in this publication:

"The National Electrical Code Committee notes the following errors in the 1990 edition of the *National Electrical Code,* NFPA 70":

First and second printing

1. In 310-13 insert "and" before 310-66 and delete "and 310-67".
2. In Figure 310-1, Note 1, change "Section 300-5" to "Section 710-3(b)"
3. Replace Figure 2 in Article 516 to correct dimensions.

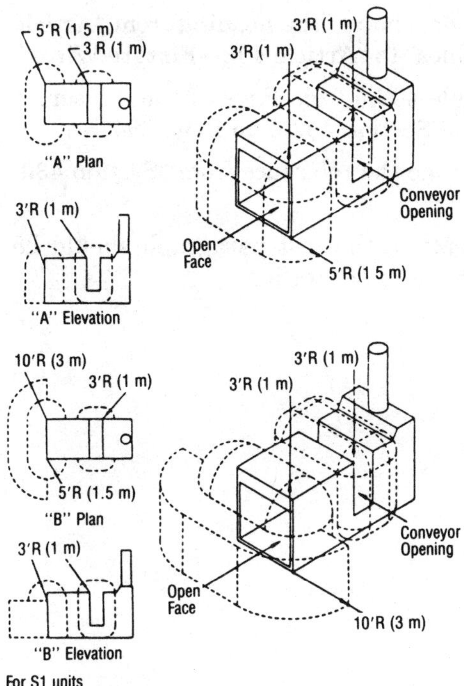

For S1 units
one inch = 25.4 millimeters.
one foot = 0.3048 meter

4. In the Index under "Guest rooms, outlets," change the reference from "210-63(b)" to "210-60".

First printing only

5. In the Committee List, under Panel No. 8, add "Table 10" to list of articles and tables.
6. In Table 220-30, revise the first line to read "Largest of the following five selections."
7. Change the first sentence in 305-6 from "...sites shall be provided as follows:" to "...sites shall be provided to comply with (a) or (b) below."
8. In Table 318-10, Footnote, change "MCM" to "kcmil".
9. In Section 370-13(e), second line, change "(1640 m^3)" to "(1640 cm^3)".
10. On page 295 of the bound code, change the heading from "Article 400—Flexible Cords and Cables" to "Article 402—Fixture Wires".
11. On page 297 of the bound code, change the heading from "Article 400—Flexible Cords and Cables" to "Article 402—Fixture Wires".
12. On page 299 of the bound code, change the heading from "Article 400—Flexible Cords and Cables" to "Article 402—Fixture Wires".
13. In the bound code, move Table 402-5 from Page 301 and insert it following the first sentence of Section 402-5 on Page 294.
14. In Section 430-2, Part C, change the reference from "Section 430-1" to "Section 430-31".
15. In Section 501-5(b)(2), Exception, the last paragraph should be part of the Exception and should be in italics.

TIAs Issued by the NFPA's Standards Council

The Standards Council of the NFPA voted to adopt two Tentative Interim Amendments (TIAs) to the 1990 edition of the National Electrical Code (NEC). This action was taken at the Standards Council Meeting held April 24–25, 1990 at Incline Village, Nevada.

The first TIA deals with the new requirement in the 1990 NEC calling for the use of hospital-grade receptacles at general care patient bed locations as covered by Sec. 517-18(b). The original wording submitted for this TIA would have permitted the use of receptacles meeting a specific Federal Specification instead of the "hospital-grade" receptacles, which are required to be used in such locations after January 1, 1991. Although the Council rejected the wording of this TIA, it did vote to issue a TIA that would extend the effective date for compliance from January 1, 1991 to January 1, 1993.

The second TIA adopted by the Standards Council involved the "Construction Specifications" given in Part C of Article 370. In Sec. 370-20(b), which regulates the minimum wall-thickness of metal boxes, an additional sentence was added to recognize zinc-plated boxes with slightly thinner walls. This new sentence, which is to be added immediately before the last sentence, reads: "The wall of a listed zinc die-cast or permanent mold cast box shall not be less than 3/64 inch (1.19mm) thick."

Proposed TIA calls attention to markings on breakers used in corner-grounded delta systems

The new Fine Print Note (FPN) following Sec. 240-83(e) in the 1990 NEC gives application information relative to the markings on circuit breakers. This FPN indicates that CBs marked by the manufacturer with just a voltage value (e.g., 480V) may be used on grounded or ungrounded electrical systems. And those CBs marked with "a slash" voltage value (e.g., 480Y/277V) may only be used on neutral grounded systems. The proposed Tentative Interim Amendment (TIA), which was submitted by the National Electrical Manufacturers Association, is intended to provide additional information in the FPN of Sec. 240-83(e) for the markings required on CBs used in corner-grounded delta systems.

As reported in the Oct/Nov edition of the "NFPA Standards Action," the proposal for this TIA (TIA No. 317) reads as follows:

> 1. Retain the existing FPN text, adding the following as a new second sentence:
>
> "Two pole circuit breakers which are suitable for controlling three-phase, corner-grounded delta circuits are investigated and marked to indicate their suitability."
>
> Submitter's Reason: NEMA believes that the current text of the fine print note is misleading and could lead to the unsafe application of certain types of molded case circuit breakers.
>
> The proposed clarification will ensure that the reader realizes that, regardless of voltage marking, when a two-pole breaker is to be used to control a three-phase circuit, only a two-pole breaker that has been specifically tested and marked for the purpose should be utilized. Because not all two-pole breakers are suitable for use in controlling three-phase circuits, this explanatory information could help to avoid an inadvertent, and unsafe, misapplication.

Whether or not the TIA is accepted, all design and installation people should be aware of the distinct marking provided on two-pole breakers that may be used in corner-grounded systems. A statement similar to the proposed wording of this TIA appears on the top of page 19 in the UL's *General Information Directory* (a.k.a., the White Book). It reads:

> Two-pole circuit breakers which are marked to indicate use on 3-phase circuits have been investigated and found suitable for controlling 3-phase, end-grounded delta circuits.

Basically stated, this UL listing instruction is indicating that two-pole breakers that are suitable for use on corner-grounded delta electrical systems will bear a marking of "1ϕ–3ϕ." Future editions of the White Book will contain this specific marking information.

Because UL already states that only two-pole breakers so marked are tested and intended for use on corner-grounded delta systems, use of other two-pole breakers on corner-grounded systems is a violation of Sec. 110-3(b), which requires listed or labeled equipment to be used in accordance with any instructions given as part of the listing. Therefore, always verify the presence of the "1ϕ–3ϕ" marking on any two-pole breaker used in corner-grounded delta circuits.

New UL Standard (UL 1950) for "Information Technology Equipment" to Replace UL 478, "Data Processing Equipment, Electronic"

On March 15, 1989, the Underwriters Laboratories issued a new standard for "Information Technology Equipment" known as UL 1950. Based on the International Electrotechnical Commission's (IEC's) Standard 950, the new UL 1950 will cover the same type of equipment previously tested in accordance with UL 478, "Data Processing Equipment, Electronic," that is, mainframes, PCs, computer terminals, other computer peripherals, and certain computer power equipment.

The main reason for UL's adoption of this standard is compatibility. Even though the fifth edition of UL 478 was almost complete, it was deemed more prudent to move toward harmonizing the standards used in Europe with those used in the United States. To that end, instead of adopting a newly revised UL standard, a modified version of IEC 950 was developed and issued. This will reduce the burden on equipment manufacturers by eliminating the need for certification by two different bodies (the IEC and UL) and the expense associated with dual certification.

Although UL 1950 is intended to replace UL 478, "Data Processing Equipment, Electronic," it does not become effective until March 15, 1992. However, any equipment submitted for listing as "Data Processing Equipment, Electronic" prior to that date *will be* evaluated as stipulated in the new UL 1950 and listed as "Information Technology Equipment," unless the equipment manufacturer requests otherwise. Equipment tested in accordance with UL 1950 will be marked as so tested.

It is important for designers, installers, and inspectors to note that equipment listed in accordance with UL 1950 as "Information Technology Equipment" satisfies the requirement given in various sections of Article 645 that calls for the use of "listed electronic computer/data processing equipment." That is, as it now stands, equipment listed in accordance with either UL 478 or the new UL 1950 *is* "listed electronic computer/data processing equipment" and may be used in "Electronic Computer/Data Processing" rooms, as covered in NEC Article 645.

Index